Latin American Development: Geogr

Latin American Development:

Geographical Perspectives

Edited by David Preston

Longman

Longman
Longman Group Limited
Longman House, Burnt Mill, Harlow
Essex CM20 2JE, England
and Associated Companies throughout the world

First published 1987
Second edition 1996

British Library Cataloguing in Publication Data
A catalogue entry for this title is available from the British Library.

ISBN 0-582-23695-9

Library of Congress Cataloging-in-Publication data
A catalog entry for this title is available from the Library of Congress.

Set by 7 in 10/11 Palatino
Produced by Longman Singapore Publishers (Pte) Ltd.
Printed in Singapore

Contents

CONTENTS

List of Figures

List of Plates

List of Contributors

Dr Anthony J Bebbington, Research Associate, Sustainable Agriculture Programme, International Institute for Environment and Development

Dr John Dickenson, Senior Lecturer in Geography, University of Liverpool

Dr Peter Furley, Senior Lecturer in Geography, University of Edinburgh

Professor Alan Gilbert, Professor of Geography, University College and Institute of Latin American Studies, London

Dr Robert N Gwynne, Reader in Geography, University of Birmingham

Dr Ignacio Irarrázaval, Centro de Estudios Públicos, Santiago, Chile

Professor Arthur Morris, Professor of Geography at the Department of Geography and Topographical Science, University of Glasgow

Professor Linda Newson, Professor of Geography, King's College, London

Dr David Preston, Senior Lecturer in Geography, University of Leeds

Dr Sarah Radcliffe, Lecturer in Geography, University of London, Royal Holloway

Dr Joseph Scarpaci, Associate Professor, Department of Urban Affairs and Planning, Virginia Polytechnic and State University

Acknowledgements

We are grateful to the following for permission to reproduce copyright material:

South American Pictures, Woodbridge, Suffolk, England, for Plates 2.1, 3.1, 4.1, 4.2, 5.1, 7.2 and 10.2; Longman Group Ltd, Harlow, England, for Figures 9.2, 9.3 and 9.4 from Gwynne, R.N. *New Horizons?* (1990); John Wiley & Sons Ltd, London, for Figure 9.1 from Gibb, R. and Michalak, W. (Eds) *Continental Trading Blocs* (1994); Scientific American Inc. for Figure 2.2 from Coe, M.D. *The Chinampas of Mexico* (July 1964); CEPAL, Santiago, for Tables 8.2, 8.4 and 8.6; Oxford University Press, New York, for Table 8.5 from *World Development Report* (1979–1993). Oxford University Press/United Nations Development Programme (UNDP) for Tables 5.5 and 5.6 from *Human Development Reports 1991 and 1994*; Routledge/International Institute for Environment and Development (IIED) for Tables 5.2 and 5.4 from *Searching for Land Reform in Latin America* (Thiesenhusen ed., 1989); Routledge/Overseas Development Institute, London for Fig. 5.1 from *Latin American NGO Series* (Fig. 1.1).

While every effort has been made to trace the owners of copyright material, in some cases this has not proved possible, and we would like to take this opportunity to apologise to any copyright holders whose rights we may have unwittingly infringed.

CHAPTER 1

Themes in contemporary Latin America

David Preston

Latin America in the world

Two major periods during the past five centuries altered the appearance of some parts of Latin America:

1. In the sixteenth century the continent was invaded and many areas were transformed by European people, predominantly from Spain and Portugal. They conquered the native inhabitants, many of whom were killed by their diseases. They used land and mineral resources to create wealth for themselves and for the European powers to whom they were either economically or politically subject. They enslaved millions of Africans to work for them and they and their descendants dominate the human population of parts of the continent and the Caribbean.
2. At the end of the nineteenth century some parts of the continent were again invaded by Europeans. This time they were settlers seeking a better life than in their homeland. They transformed the economies of the southern countries of South America: they peopled the growing cities: they also helped to establish financial and other service institutions which laid the foundation for the growth of modern economies.

Even so, parts of the continent remained in the hands of the non-European original inhabitants. In these areas the landscape and human population remained substantially unchanged, and economy and ethnic identity were distinct even though changes continued to modify details of individuals, households and communities. Such areas were mainly rural, far from towns and cities. Native people lived in much of the central Andes, parts of highland Mexico and Central America and in the Amazon basin and forested areas of north-west South America as well as in some arid and semi-arid lowlands from Argentina to Mexico.

European domination?

Those parts of the continent of Latin America whose economy and population is predominantly Hispanic or of mixed race have long been a part of the European economic and cultural world. The European origins of many Latin Americans and the European influence of many of the economic and cultural developments during the nineteenth century and up to the Second World War made this inevitable. Native people were relatively few in number in some parts of the continent, they had little political power and were often seen by the ruling classes as deeply conservative or opposed to change. This ensured that they continued to be subject to discrimination or even genocide, for they had few rights and limited access to power. After the success of the various independence movements of the early nineteenth century nationalism became an important political force in many countries. Areas that, to the ruling élite, were apparently empty and under-used were incorporated into national territories without regard for the native inhabitants. The haphazard boundaries of the new states of the 1820s and 1830s were modified by both war and treaty.

Any account of Latin America as a distinctive economic or cultural realm must recognize three levels of focus:

1. One domestic and local in which individuals and groups develop their strategies for reproduction and self-fulfilment and to defend themselves against external pressures.
2. One national in which local and regional economic and social changes create a national identity.
3. The other global in which events in Latin America must be seen as part of a global pattern.

Global influence

The French Revolution of 1789 is a necessary part of the backdrop of the independence movements of Latin America just as the world boom in vegetable oil demand fuelled the dramatic expansion of soyabean production in Brazil, Argentina and Mexico in the 1970s and 1980s. Neither national nor agricultural land use revolutions took place without influence from other countries nor the world trading system.

During the period in which Spain and Portugal dominated the Latin American externally focused economy, the basis of wealth for the Spanish and Portuguese was primary products from mines and agricultural areas. The flow of gold and silver to Spain was vital for the Crown which was chronically indebted. Much silver also entered European and East Asian economies by legal and illegal means through the Philippines, Brazil and to the British and the Dutch through the Caribbean. The precious metals provided earnings which were partly spent on fine European and Asian manufactured goods – textiles, for instance – and thus the Spanish part of the Americas was an integral part of the world trading system of the sixteenth, seventeenth and eighteenth centuries. Once manufacturing on a large scale became important in Western Europe and later north-eastern Anglo-America, an increasing demand for food stimulated production in Central and South America. By the end of the nineteenth century technology allowed the rapid shipment to Europe and Anglo-America of perishable produce such as meat and bananas from places such as Costa Rica, Guatemala, Uruguay and Argentina. Thus some areas which had produced little surplus during the colonial period became incorporated into world trade.

Although our focus is geographical – Latin America – it is necessary to recognize that the growth and decline of specific regional economies in many parts of Latin America are closely linked to events outside the continent. The demand for and prices of a commodity produced in Brazil for sale in Germany or Illinois is a function of the nature of the economic system which affects those goods and not just the conditions in the area of production. It is logical to recognize the value of using a world-systems perspective in seeking to understand some of the historical and geographical patterns of change in Latin America (see Box 1.1).

Box 1.1 Wallerstein's world-systems

The development and refinement of the notion of world-systems as a necessary tool for the analysis of social behaviour has been associated with historian Immanuel Wallerstein. He is particularly involved in the description and analysis of one particular world-system: the capitalist world-economy. He identifies three characteristics of a world-systems perspective: one is the use of the world-system as the unit of analysis for the study of social behaviour; another is the recognition of the importance of the long duration of certain periods of note in human history; and the third is the view of the capitalist world-economy.

This approach offers a distinctive framework for historical analysis, it stresses the importance of using the knowledge gained from the work of scholars in different disciplines, it stresses the dynamic nature of economic and social structures which makes the indentification of beginnings and ends of notable periods difficult. World-systems analysis also stresses the importance of recognizing the changing nature of spatial units in response to different perceptions and geographical levels of resolution.

Source: Wallerstein (1990)

Latin America is one of the more industrialized and affluent continents among those that are peripheral – that is, relatively powerless – in the world economy which is dominated by the USA, Japan and the European Union (EU). In the world trade negotiations at the beginning of the 1990s it finds itself split between the possibilities of two geographical alliances. One is with the USA, which is particularly attractive for Mexico which shares a long common border with the USA and the other, for certain countries, is with the so-called Cairns group of countries heavily dependent on the export of agricultural commodities a surplus of which is produced by both the USA and the EU.

A question of scale

One aim of this book is to help the reader grasp the meaning of trends and events to people and places in Latin America. The emphasis of the previous paragraphs has been on the continent as part of a global system from which it cannot completely escape. Even so, as Bob Gwynne explains in Chapter 9, economic change has taken place sometimes independent of stimulus from the world-economy. It is necessary to understand that the functioning of the capitalist world-economy affects Latin America at every imaginable scale from continent to village and down to household.

A. G. Frank suggested that there was

> a whole chain of metropolises and satellites, which runs from the world metropolis down to the hacienda or rural merchants who are satellites of the local commercial metropolitan centre but who in their turn have peasants as their satellites (Frank 1967: 146–7).

Thus one can readily imagine that many aspects of Mexico are powerfully influenced (dominated) by the USA, both economically and culturally. At a national level it is easy to understand how actions by the US government to release silver on the world market, or to check the movement of migrants across its borders with Mexico can hardly be resisted by a Mexican government of any political complexion. The consequences of such actions affect not only the national economy but thousands of Mexicans of different social classes and in many parts of the country.

The members of the Sánchez household in Mexico City are poor and find a range of ways to make a living. Manuel is attracted by the opportunities of working in the USA and travels there illegally. The relative ease of access, the rates of pay and type and duration of work

---

Box 1.2   Consuelo, talking about her family

if someone gave Manuel [her brother] a common stone, he would hold it in his hand and look at it eagerly. In a few seconds, it would begin to shine and he would see that it was made of silver, then of gold, then of the most precious things imaginable, until the glitter died.

Roberto would hold the same stone and would murmur, 'Mmmm. What is this good for?' But he wouldn't know the answer.

Marta would hold it in her hand for just a moment, and without a thought, would throw it carelessly away.

I, Consuelo, would look at it wonderingly, 'What might this be? Is it, could it be, what I have been looking for?'

But my father would take the stone and set it on the ground. He would look for another and put it on top of the first one, then another and another, until no matter how long it took, he had finally turned it into a house.

Source: Lewis (1963: 274)

---

there are all influenced by the state of the US economy and of the economy of California where he seeks work. His father, Jesus, has a steady job in Mexico City working in a restaurant owned by a Spaniard. His daughter Consuelo has a range of jobs, mostly short-term, depending on the dynamism of the city's economic growth and her own physical strength (see Box 1.2). One household is thus influenced to varying degrees by events in different parts of two countries as well as by the state of the city's economy (consult Oscar Lewis's (1963) *Children of Sánchez* to come to your own conclusions and see Box 1.3).

Box 1.3 Oscar Lewis and Latin American people

The work of Oscar Lewis, a US anthropologist, is a valuable source of knowledge about the lives of the poor in Latin America. He spent most of his career, from 1943 to 1970, studying the lives of the poor in Mexico, New York City and Havana. He was admired by both *Time* magazine and Fidel Castro. Many middle-class Latin Americans were outraged that so much publicity should be gained by a North American professor portraying the lives of some of the less privileged members of their society.

He worked by tape-recording the life histories of individuals, and his books were largely the edited and translated autobiographies of his subjects. As such, they offer an outstanding opportunity for people interested in Latin America to read about the lives of some of its people.

Scholars have made little use of his work although generations of students have enjoyed reading some of his books.

The example of the Sánchez household shows that it is necessary to recognize both *dependence* and *independence* as important elements that can help us understand two fundamental aspects of contemporary Latin America:

1. The crucial links between past and present events.
2. The relationships between people and between geographical locations.

Elaborating on A. G. Frank's view of the levels at which dependence can exist in relation to Mexico, the different geographical and domestic scales can easily be shown. Mexico is dominated by the USA in many ways. In 1995 Mexico was forced to contract payments of her oil revenues to the USA in return for help in tackling the economic crisis of 1995. Needed foreign investment would not have returned without giving in to powerful US and world financial pressures.

Elaborating on A. G. Frank's view of the levels at which dependence can exist, a commentary is given below in respect of four different levels at which there is a dependent relationship between geographical locations: the country, the capital city, the regional capital and the village.

- Within Mexico, Mexico City dominates economic and social development. A university in Oaxaca will be able to attract lower-quality professors than one in the capital. The city and state of Oaxaca will have less money to spend on education, health and roads because most of the population lives in the countryside and produces mostly for their own subsistence.
- The village of Tequi two hours from Oaxaca has difficulty recruiting and keeping teachers for its 170-pupil primary school, for the housing is poor compared with that in the city and the village is relatively isolated: only one bus a day. Many decisions about the running of the school and about the provision of facilities in the village are controlled by bureaucrats in Oaxaca.
- Within the village the Martínez household recently set up home in the settlement having moved from an isolated site over an hour's walk away. The eldest son has been working in Mexico City for a year, and his sister who accompanied him on his last trip to the capital, has a job in a restaurant. The money that they could send home enabled their ageing parents to move to Tequi, start a small shop and also help their younger siblings find work to supplement their earnings from farming their land. One or two of the household usually live in the old home an hour away.
- The hamlet of Kenchi, an hour on foot from Tequi, has only one teacher in its single-room thatched school, water is taken from nearby streams and the one house that sells goods often runs out of basic items such as matches and cooking oil because the woman who runs the shop does not have the money to buy much when she goes to market. The people there are despised by those in Tequi for their relative poverty and for the facts that everyone speaks the local indian language most of the time and that they sell little in the local market.

Each household, institution or settlement in the series identified suffers from discrimination from that to which it is subordinate – its own growth and change is dependent on, and partly controlled by – decisions and actions taken by people and groups in the higher-order place. The range of geographical scales at which dependency operates needs to be clearly recognized.

Dependency is a useful concept but it has its limitations. It implies the superiority of North over South and of global economic might over local or regional power, which is not always true and demeans those in the South. It is also possible to fight against dependency. This can be done by using local knowledge and skills to produce what previously was imported from elsewhere, by developing links with still higher-order settlements and playing one more powerful partner off against another to get a better deal, and by direct action such as destroying power lines, blocking highways and diverting water destined for urban centres to highlight local needs. Local and regional individuals and groups have frequently acted independently and with success to create new structures which aid their survival or facilitate success.

The globalization of culture

The ability of modern communications to allow information to be transmitted and transformed to people who have few means to move to metropolitan centres is a major change which has affected regional and national culture in most parts of the world and throughout Latin America during the past 40 years. Now, in most small towns and even large villages some households have television sets and almost all have radios. Major world events and all national sporting events become a matter for comment within hours of their occurrence. In villages without electricity vehicle batteries provide power for televisions and video recorders. Even photocopiers are available in small towns which likewise enable local and national information to be widely and quickly circulated.

The widespread migration between and within rural and urban areas for long and short periods further allows people to be aware of other ways of life and means of obtaining income or gaining power. Much migration in some countries is across national borders and knowledge of other nations is likewise easily obtained. Returning migrants do much to spread news of the similarities and differences between their homeland and the places to which they move.

The ready availability of cheap manufactured goods imported from other parts of the world, in particular East Asia, makes it economically possible for a Brazilian farm worker to buy blue jeans made in Taiwan and for his wife to have an Orlon jumper knitted in Rio with fibre from Hong Kong. The range of radio stations in cities and larger towns allow the diffusion of popular music from other countries, while cassettes enable such music to be played in bedrooms and bars.

Paradoxically such media have made it possible for major musical trend-setters to popularize music from Latin America (Paul Simon) and the emigration of Latin Americans to big cities in the North has stimulated awareness of Latin music played by Latin Americans. Market-places in Chicago and Hamburg, Paris and San Francisco feature musicians from Andean countries playing to interested audiences.

The globalization of new cultural trends also leads to a heightened consciousness of the differences in life-styles and levels of living in different parts of both countries and continents. This makes poor and marginalized people increasingly aware of their position in society and also the limited improvement of their quality of life. On the other hand, regional and indigenous cultures also flourish alongside contemporary youth culture influenced by Northern values and symbols. National and regional artists – musicians, sculptors, film-makers and writers – are greatly respected both at home and overseas. The same media which enable a greater awareness of Northern cultural forms also allow the diffusion of local and regional culture. Even so, Northern cultural forms are more effectively promoted by advertising and dominate cinema, film and television.

Latin American identity

Ethnic identity

The native people of Latin America are far from homogeneous even if all are descended from the people who crossed into the Americas via the Bering land bridge as long as 30,000 years ago. There were contacts through trade, warfare and exploration between forest people and mountain dwellers and between Mexico and Central America and the southern continent long before the arrival of the Spanish. The local and regional cultural forms that had evolved by 1500 were not eradicated by conquest. Many still exist in a twentieth-century form among people who prefer to speak their own language, cherish their role as part of a vibrant and dynamic social group with many different forms of expressions of cultural identity. The many rebellions of native people against white oppression and forms of exploitation were occasions when local and regional ethnic individuality was less important than the solidarity of the native people against the dominant outsiders. It therefore logical that they are still conscious of sharing a common set of problems because of the discrimination practised against them by members of the dominant groups in regional and national society. The many forms of popular protest against national and international abuse used by native American groups in the 1990s shows that their identity has been heightened rather than eroded.

Afro-Americans have a separate identity from indigenous people and, from the colonial period, their religion and music have been recognized as distinctive, as necessary for their collective well-being and, also, of value for white people. A large number of Afro-Americans existed in some parts of Latin America, whether as slaves, free people and deliberate 'outsiders' in Maroon societies of ex-slaves. This has made the identity of black people as important a regional and national issue as that of native Americans in other areas.

Assimilation of immigrants into new countries and 'new' cities

The immigrants to Latin America from the overcrowded cities of Europe and poverty-stricken rural areas in the Mediterranean basin, East or South Asia found a society into which they could insert themselves. Although they were discriminated against as latecomers they brought with them capital and energy that was often in short supply. Above all, they brought a diversity of cultural values, knowledge and skills. Although many were dominated by families who had arrived centuries before on whose land they initially worked, there were opportunities to start new businesses in rapidly growing cities, to work for other immigrants in new enterprises or to move to the fringes of white settlement and start farming in areas inhabited by small numbers of indigenous people.

The process of incorporation of many of the immigrants was similar to that in the North. Early segregation gave way to gradual assimilation.

Some groups came to escape discrimination and were encouraged to settle the outermost parts of countries such as Paraguay or Brazil. They formed distinctive settlements and often specialized in particular crops or crafts as well as maintaining their cultural identity. However, within a few generations the children and grandchildren of the original settlers had become less culturally distinctive and began to merge into whatever social class their economic status had demonstrated entitlement. This was as true of the Welsh in Patagonia as the Mennonites in Paraguay or Belize or even the Chinese in Peru.

Regional and hemispheric identity through alliances and political negotiations

Latin American countries have formed few regional alliances that have had any lasting impact on hemispheric relations. The Organization of American States is controlled and dominated by the USA and has shown little skill in allowing Latin American voices to be heard on a world stage. Even in the nineteenth century the unions of Greater Colombia (Ecuador, Colombia and Venezuela) and Central America (Guatemala, Honduras, El Salvador, Nicaragua and Costa Rica) lasted less than a decade and were incapable of showing their importance to avoid the dominance of either the USA or the major Latin American nations such as Mexico, Brazil, Argentina or Chile.

The dominance of one commodity in export earnings for long periods made such countries vulnerable to economic blackmail from the major importing nations, particularly the USA and formerly the UK. The production of similar export commodities made competition more natural than collaboration and the rise of African and other Southern producers often made even Latin American cartels powerless in the face of even disunited competition.

The interests of the USA during much of the twentieth century dominated relations between Latin American countries. The fear of communism which dominated the four decades after 1945 made individual national attempts to redefine their role in the world subject to enormous pressure to conform from the USA. Thus the major political revolutions in Latin America have mostly succumbed to pressure and have had only limited symbolic effect in Latin America. Nicaragua and Chile both changed their political complexion after intense support for opposition from US sources and only Cuba and Mexico have survived without opposition-led change. Even so the Mexican Revolution has long since been revised to serve economic and political interests within Mexico that conform to ideologies widely supported in the USA.

Cuba remains as a symbol of what can be achieved by a single nation, originally favoured by imaginative revolutionary leadership, and supported by a major global power. Cuba has had immense symbolic importance in Latin America, in particular for the young. Its major achievements in health, education and sport are well recognized throughout the world. Yet it failed to spread its doctrine beyond university campuses and labour unions to stimulate imitation by an Latin American state save Nicaragua, under special circumstances. It had more influence among the young nations of southern Africa through its involvement in liberation struggles in Mozambique, Namibia and Angola than among Latin American nations.

Individualism rather than regional solidarity marks the Latin American political scene. This is often linked with the continuance of territorial disputes between adjacent states and the exaggeration of these issues by military governments or military interests within states.

Comparing alliances in Latin America with South-East Asia

The other part of the South with which Latin America should be compared is the group of nations of South-East and East Asia some of which have changed even more rapidly than Latin America in the post-1945 period. Many nations of eastern Asia also experienced colonial domination by the British (parts of China, Malaysia), Spanish (Philippines), French (Vietnam and Cambodia) and Dutch (Indonesia). Although many were also affected by the war with Japan, they also became gradually incorporated into the world economy during the present century. To test whether any search for a Latin American identity is really realistic, it is useful to review the experience of eastern Asia in the emergence of any regional identity.

Further reading

Lewis O (1962) *Five Families*. Wiley, New York. (An account of a day in the lives of five families, each of which is the subject of a further book-length study.)
Reck G (1978) *In the Shadow of Tlaloc: Life in a Mexican Village*. Penguin, Harmondsworth. (A moving, personal account of all aspects of life in a small Mexican village.)

CHAPTER 2

The Latin American colonial experience

Linda A. Newson

The separation of the Old and New Worlds ended in 1492 with Columbus' landfall in the Bahamas. It also heralded the emergence of a world economy that would eventually be truly global. Columbus encountered a continent that had been settled for maybe 30,000 years and probably contained at least 50 million people. During the pre-Columbian period native cultures had diversified such that on the eve of European expansion Latin America was occupied by a mosaic of societies ranging from simple hunting, fishing and gathering bands to highly developed agricultural states. The colonial period that followed saw the incorporation of these societies not only into the Spanish and Portuguese empires, contributing to Latin America's distinctive character, but also into the wider world economy. These processes brought a degree of cultural uniformity to Latin American societies, but regional variations persisted which reflected not only the priorities and policies of the colonial powers, but also the continent's diverse environments and its pre-Columbian heritage. As such, it is impossible to understand developments during the colonial period without a knowledge of both the character of native societies on the eve of European expansion and an understanding of the cultural environment in Europe that led to the conquest of the New World.

The pre-Columbian heritage

The first inhabitants of America were hunters and gatherers who arrived from Asia having crossed the Bering Strait during the last glaciation. Since the earliest human settlers were few in number and were nomadic hunters, fishers and gatherers, they left little trace of their presence.

From 8500 BC a wide range of plants was domesticated, including maize, potatoes and manioc, which after 1492 spread world-wide and today figure among the world's major staples. Other domesticates were beans, pumpkins and squashes, peanuts, sweet potatoes, fruits such as the pineapple, tomato and avocado, chilli peppers, tobacco and cacao. The earliest evidence of domestication comes from Guitarrero Cave in the Peruvian highlands, where common beans, lima beans and chilli peppers date from about 8500 BC. In Mexico maize was being cultivated at Tehuacán about 5000 BC, while in Oaxaca cultivated bottle gourds and pumpkins have been found dating from about 7400 BC. These crops have formed the basis of different farming traditions in the New World, with farmers in Mexico concentrating on the cultivation of maize, beans and squashes (pumpkins) and those in the Andes on potatoes and quinoa, and at lower elevations sweet potatoes and manioc. Unlike the Old World, few animals were domesticated in the Americas. In the Andes the domestication of the llama, alpaca and guinea pig (cavy) began about 4500 BC, and in Mexico turkeys and ducks were raised and dogs were kept. Domestication in the New World is clearly as old as that in the Old World. However, differences in the crops domesticated and the manner of sowing (the planting of single seeds in the New World as opposed to broadcasting in the Old), the relative absence in the New World of domesticated animals and of significant technical innovations such as the wheel and iron, argue for independent developments in the two areas.

Even though domestication may have begun about 8500 BC, it was a slow and cumulative process and several thousand years elapsed before agriculture formed the basis of native subsistence. In Mexico farming villages did not appear until about 2500 BC. Sometimes permanent settlements preceded the development of a fully fledged agricultural economy. This was the case on the coast of Peru and the Gulf Coast where more reliable aquatic resources enabled villages to emerge about 3000 BC. As farming became more common, food supplies were more abundant and assured. This enabled groups to become sedentary and permitted larger population densities to be supported.

At first agriculture was land-extensive, involving the periodic abandonment of plots to fallow to allow the soil to recover some of its lost fertility. This type of cultivation, known as swidden or shifting cultivation, was once widespread in Latin America, but in many areas it was later superseded by more intensive forms of production. At the time of the Iberian conquest it persisted mainly in the tropical forested lowlands, notably Amazonia, where it was combined with hunting, fishing and gathering. It supported kin-based communities of up to several thousand people, though they remained essentially egalitarian.

By 1492 new techniques which intensified and extended agricultural production had been introduced to many areas. They included techniques of irrigation and terracing, particularly common in the Andean area, the use of fertilizers, and the construction of raised fields and floating gardens, known in Mexico as *chinampas* (Box 2.1 and Figs 2.1 and 2.2).

1. Irrigation, often combined with terracing, not only extended production in semi-arid areas, but helped maintain soil fertility by supplying it with water-borne nutrients. It varied from simple pot irrigation to more sophisticated canals and aqueducts which were most highly developed on the north coast of Peru.

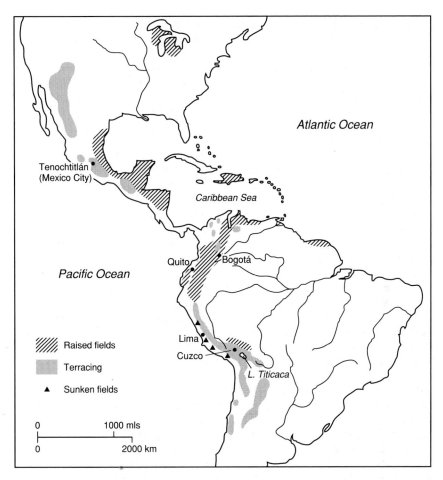

Fig. 2.1 Pre-Columbian forms of intensive agriculture (*Source*: based partly on Donkin 1979)

2. Terracing not only enabled the cultivation of steep slopes, but helped maintain soil fertility by discouraging soil erosion and improving water retention and absorption. In Peru at least 600,000 hectares of terracing exist, most of which are prehistoric in origin (Denevan 1992a).
3. Less familiar, but almost equally extensive, were drained and raised fields designed to reduce waterlogging, and thereby also improve soil temperature and aeration. These were constructed in moist tropical lowlands, such as the Guayas basin of Ecuador, around highland lakes, such as in the Valley of Mexico and around Lake Titicaca, and on the slopes of the northern Andes.

These techniques not only expanded agricultural production and enabled larger populations to be supported, but called for a greater

Box 2.1 *Chinampas* or raised fields

Chinampas were raised platforms approximately 30 metres by 2.5 metres constructed of layers of reeds and mud, the edges of which were anchored in place by trees. Not only were the fields raised above the water table thereby enabling cultivation, but organic and nitrogenous material could be dredged from adjacent ditches and canals to enhance soil fertility and allow continuous cropping. In addition, fish and other aquatic life might be exploited. It has been estimated that the productivity of *chinampas* may have been five times higher than that of conventional cultivated fields. Each property consisted of a house and about six or eight *chinampa* strips. About 12,000 hectares of *chinampas* were constructed around the Aztec capital, Tenochtitlán, whose population on the eve of the Spanish conquest had reached some 250,000.

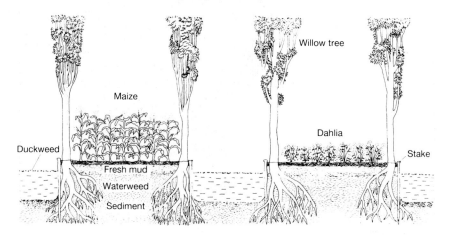

Fig. 2.2   Cross-section diagram of *chinampas* and canals gives an idea of their construction. Fresh mud from bottom of canals and weeds for compost beneath the mud keep the *chinampas* fertile. Trees and stakes hold the sides of the *chinampas* firmly in place (*Source*: based on Coe 1964)

degree of social and political organization. Chiefdoms and states emerged with populations exceeding tens of thousands distributed in hierarchies of settlements ranging from small villages to large urban centres. These societies were politically organized and socially stratified, possessing leaders who had the authority to exact tribute and labour services from commoners. By the time of the Iberian conquest several civilizations, such as the Maya of Yucatán and Guatemala and the Tiwanaku in highland Bolivia, had collapsed, but the Aztec and Inca empires were still in the process of expansion (Fig. 2.3).

Considerable controversy exists over the precise size of their aboriginal populations. This derives from the fragmentary nature of the archaeological and historical evidence, differences in the methods of estimation employed by researchers and their views on the level of

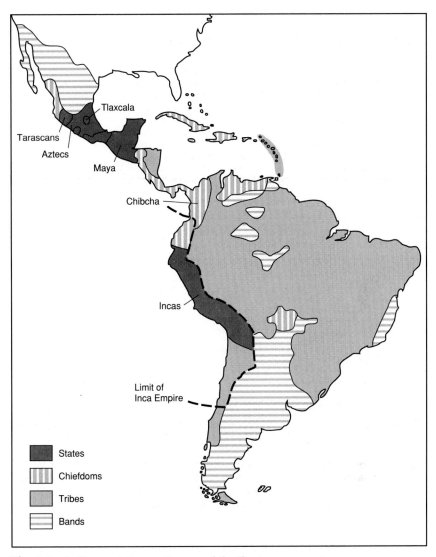

Fig. 2.3 Indian societies on the eve of the Iberian conquest

depopulation attributable to disease and colonial rule. Denevan's (1992b) review of estimates of the aboriginal population of the Americas suggests that Latin America contained about 50 million people, but others, such as Dobyns (1966), would double this figure. The native population was concentrated in the highlands of central and southern Mexico and the Andes. Borah and S. F. Cook (1963) have estimated that the population of central Mexico in 1492 was about 25 million. Although some authors would reduce this figure by more than half, it seems probable that the population of central Mexico exceeded that of the

Andean area; N. D. Cook (1981) has estimated that the aboriginal population of Peru was about 9 million. Other large concentrations of Indian population were to be found in Central America, Colombia, Ecuador and Bolivia, the most notable of which were the Maya of the Yucatán peninsula and Guatemala, and the Chibcha of Colombia. Compared with these areas, the rest of Latin America was sparsely settled, with the vast Amazonian area containing maybe only 5 million people (Table 2.1).

Table 2.1 Aboriginal population estimates

	Dobyns (1966)	Denevan (1992b)	Denevan (1992b) (per cent of total)
Mexico	30,000,000–37,000,000	17,174,000	34.3
Central America	10,800,000–13,500,000	5,625,000	11.2
Caribbean	443,000–553,750	3,000,000	6.0
Andes	30,000,000–37,500,000	15,696,000	31.3
Amazonia		5,664,000	11.3
Chile, Argentina		1,900,000	3.8
Paraguay, Uruguay, and southern Brazil	9,000,000–11,250,000	1,055,000	2.1
Total	80,243,000–100,303,750	50,114,000	

Source: based on Dobyns (1966) p. 415 and Denevan (1992b) p. xxviii

It is worthy of note that the distribution of native population in Latin America contrasted markedly with that found in eastern and southern Asia, where the greatest concentrations were to be found in the lowlands, rather than the highlands. In Asia the development of irrigation agriculture underpinned the emergence of civilizations in the major river valleys and deltas, whereas in the highlands tribal groups continued to practise less intensive forms of cultivation.

Not only was Latin America highly populated, but its landscape bore the heavy imprint of human activity (Denevan 1992a). From the earliest years of human occupation, the structure and species composition of its forests had been modified by fire and clearance, and grasslands extended. In some areas, such as Central Mexico, soil erosion was severe and food shortages common. Even the apparently pristine tropical forest of the Amazon basin had not escaped human interference, as evidenced by the extensive underlying deposits of black soil or *terra preta*, containing charcoal and sometimes pottery and other artefacts. More obviously the land had been transformed by terracing, the construction of raised fields, irrigation canals, causeways, roads and settlements.

The character and distribution of native populations were to play significant roles in determining the nature and pattern of European conquest. Areas occupied by chiefdom and state societies were more attractive to colonists, since they contained large sedentary populations from whom tribute, often in the form of gold and silver, and labour services could be exacted. Furthermore, pre-existing political structures enabled the Spanish to bring them under colonial rule more easily. On the other hand areas occupied by tribal groups and hunting, fishing and gathering bands proved less attractive since these societies produced little in the way of surpluses or sources of labour. In addition, they were more difficult to control since they lacked strong native leaders and in the case of hunters and gatherers were nomadic.

The Spanish and Portuguese were not only interested in exacting tribute and labour services from native peoples and converting them to Christianity, but they also wished to profit from the region's natural resources, amongst which gold and silver, and later tropical products, were the most important. These resources could not be exploited, however, without labour and as such the pattern of native settlement played an important role in determining where European enterprises developed. However, minerals had to be mined where they were located, and if they were found in areas where the Indian population was sparse, such as in northern Mexico and in Minas Gerais, Brazil, labour had to be imported from other regions or from Africa.

The colonial period

In the fifteenth century three groups of people were capable of ocean navigation: the Chinese, Arabs and Europeans. Why was it then that Europeans rather than the other two groups eventually established contact with the New World?

1. The Chinese had little incentive to expand overseas since they were largely self-sufficient in luxury goods as well as daily necessities, and they were often preoccupied with the Mongol threat to their northern borders.
2. Meanwhile Arabs had explored the Indian Ocean and coast of Africa, and had established trading links with China and South-East Asia as far as Indonesia, but they were not politically unified and their ships were inferior to those built in Europe in terms of manoeuvrability and speed.
3. The Portuguese had already explored the Atlantic islands when in 1453 Constantinople (today Istanbul) fell to the Turks, effectively blocking overland routes to Asia. This not only stimulated the search for a sea passage to Asia, but, facilitated by the unification of the Crowns of Castile and Aragon, also spurred the reconquest of southern Spain from the Moors.

The expulsion of the Moors was complete in 1492 when Columbus discovered America, and in many respects the conquest and colonization of the New World can be regarded as an extension of the Reconquest: Spain with a centralized government, full of missionary zeal, and with the recent experience of conquest was ready to expand overseas. Meanwhile, the Portuguese had gained valuable overseas experience in the colonization of Madeira and the establishment of trading posts in West Africa. Even after Portugal gained a foothold on the South American mainland as a result of Pedro Alvares Cabral's expedition in 1500, her interests continued to focus on trade with Asia, and Brazil remained a neglected colony until gold was discovered there at the end of the seventeenth century.

The initial conquest and colonization of the New World was the product of co-operation between the state and private individuals. The Crowns of Spain and Portugal made contracts with conquistadores to undertake and provide financial backing for expeditions of exploration and conquest in return for which they granted them certain privileges, such as the authority to found towns and to issue land grants. Nevertheless, in this joint enterprise the state was the dominant partner and once the New World had been effectively conquered, both Iberian powers began to reassert their authority and restrict the privileges of early conquistadors and colonists. This involved the establishment of administrative bureaucracies entrusted with enforcing laws that aimed at controlling every aspect of colonial life. By 1680 no less than 600,000 laws had been passed relating to Spanish America. The administration in Spanish America was far more extensive than in Brazil. The region was divided into two viceroyalties – the Viceroyalty of New Spain and the Viceroyalty of Peru – with the boundary running across the Panamanian isthmus, and these in turn were divided into *Audiencias* presided over by presidents (Fig. 2.4). Within the jurisdiction of the *Audiencias*, governors, *corregidores* and *alcaldes mayores* were also appointed, together with a host of legal and financial officials. Many of these officials had purposely overlapping jurisdictions so that a form of internal spy system was created which kept a check on their conduct and promoted the power of the Crown.

However, colonial rule in Latin America not only involved the establishment of an administration, as it did in the case of British rule in India, but also colonization. The colonization of Latin America, however, differed from that in other settler colonies, such as North America, Australia and New Zealand, in that much of the territory was densely settled. Colonization in Latin America was therefore more complex. The prime aims of the Spanish and Portuguese Crowns in America were to 'civilize' and convert the native population, as well as to exploit it and the region's natural resources as sources of profit. Conquistadors and colonists were more exclusively interested in enhancing their wealth and improving their social positions. It is often said they went to Latin America for 'god, gold and glory'. With these objectives they focused initially on areas which were inhabited by dense Indian populations or which possessed easily worked deposits of gold. Once these quick and easy sources of profit had been plundered and exhausted, colonists turned their attention to economic activities which required greater investments of time and capital, and which generated more permanent settlements. Silver mining became the driving force of the colonial economy, whilst agriculture developed to meet European demands for tropical crops but, above all, to supply growing domestic markets in the towns and mining areas.

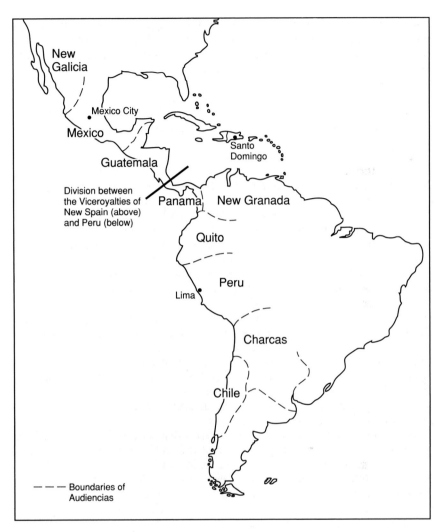

Fig. 2.4 Spanish colonial administrative divisions

Few colonists were attracted to areas which lacked minerals or could not produce agricultural products for which there a European market, or where the native population was relatively sparse. Notable amongst these were large parts of Uruguay, Argentina and southern Brazil, which produced in the main temperate agricultural products for which a European demand did not develop until the Industrial Revolution. Such regions attracted few European settlers, the most conspicuous arrivals being missionaries who were entrusted with the task of converting and 'civilizing' their small native populations, and cattle, which multiplied rapidly on the grasslands. Thus in the colonial period, it is possible to divide Latin America into core areas, where European enterprises and

19

the native population were concentrated, and peripheries which remained largely uncolonized (Fig. 2.5); the location of these cores and peripheries altered in the nineteenth century with changing demands for products in Europe.

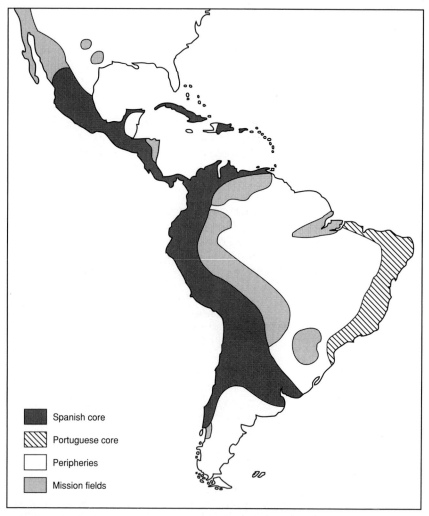

Fig. 2.5 Cores and peripheries in the late colonial period

Core areas

Urban centres
Towns and cities were primary instruments of colonization particularly in Spanish America where they were centres of political and economic power; in Brazil this role was assumed by the large estates or plantations, at least up until the second half of the eighteenth century

when the Portuguese Crown tried to enhance the prestige of the cities in an attempt to counteract the power of rural landlords. As in southern Spain during the Reconquest, towns and cities in Spanish America were employed as symbols of territorial possession and centres from which the surrounding countryside could be administered and exploited. The lowest level of administration was the town council, known in Spanish America as the *cabildo* and in Brazil as the *câmara*, which was responsible for the allocation of land grants, setting market prices, etc.; there was no form of rural government for the jurisdiction of one town extended to that of another. Spaniards were required to live in urban centres in the belief that an urban way of life was more civilized and that the residential segregation of Spaniards and Indians would protect the native population from exploitation and also discourage the emergence of powerful feudal estates. The distinction between the urban Spaniard and the rural Indian was never fully realized, however, since some Spaniards did reside on their estates, whilst Indians increasingly migrated to the towns in search of employment. Even so, power and wealth remained concentrated in the urban areas.

The most important towns in colonial Latin America were political and cultural centres. Only a few towns, such as the mining centres and major ports, performed economic roles. Mexico City and Lima, as the capitals of the two viceroyalties and possessing the seats of archbishops, universities, convents and hospitals, remained the most important cities in Latin America throughout the colonial period. Their populations were only surpassed for a brief period by the silver mining town of Potosí, which at the height of silver production in the early seventeenth century was the largest city in Latin America with an estimated population of over 160,000 inhabitants. Potosí was one of the few towns in Spanish America that performed an essentially economic role; the majority of towns were political entities that were imposed on an underdeveloped economic base. Where local economies developed to meet the demands of the European and, to a lesser extent, domestic markets, trade stimulated their growth. However, the development of regional economies with well-developed urban hierarchies was retarded by restrictions on interregional trade which persisted until 1778. Significantly, the major towns in Spanish America today remain those that in the colonial period were the major administrative and cultural centres, rather than those that performed economic roles.

The dominance of political motives in the foundation of towns and cities was also reflected in their location and form. The first towns to be established were in the Caribbean – Santo Domingo was founded in 1496, Havana in 1514 and Panama City in 1517 – and it was from these centres that the conquest of the mainland was launched. On the mainland, towns were generally founded in areas of dense native population, that is, generally in the highlands, and many were located on or near existing native settlements. In 1521 Mexico City was founded on the site of the Aztec capital of Tenochtitlán, and Cuzco, the Inca capital, was converted into an effective Spanish town in 1534. Bogotá, Quito and Santiago de Guatemala were all founded in areas of dense native population. However, the extension and perpetuation of the empire also required contact with Europe. Major ports developed on the coast for the export for minerals and tropical crops, and the importation of European manufactures. The most important were the ports at which the Spanish fleets called – Veracruz, Portobello and Cartagena – but other notable foundations on the Pacific coast were Guayaquil, Lima/Callao and Arica, the latter being the port through which the silver from Potosí was exported.

Towns in Brazil differed from those in Spanish America in that the majority were located on the coast. In Brazil the native population, being relatively sparse, was an insignificant factor in determining the location of towns, which tended to mirror the pattern that had existed in Portugal. Only with the discovery of gold in Minas Gerais in the early 1690s were towns established in the interior.

Consistent with its desire to control every aspect of colonial life, the Spanish Crown insisted that towns established in the New World were to follow a specified pattern – the grid plan – whose precise form and dimensions were codified in 1573. The centre of the town was to be marked by a central square or plaza around which public buildings, such as the governor's palace, cathedral and town hall, were to be located. These were surrounded by rectangular blocks, the most central of which were generally occupied by the town's most prominent citizens, whilst the lower classes lived on the outskirts of the town, where separate Indian suburbs were also established. Although the form of towns and cities varied in detail, the grid plan was followed with such regularity that any visitor to Spanish America today will be impressed by the monotony of the urban morphology which is evident at all levels of the urban hierarchy, from the smallest provincial town to the capital city. Exceptions to this rule were the mining towns, such as Potosí, Tegucigalpa and Taxco, which grew rapidly as irregular settlements before town plans were drawn up and they were officially established as towns. The mountainous terrain in which many mining centres were located also militated against the imposition of a regular plan. These irregularly developed towns are a welcome deviation from those generally found in Spanish America, and they are more like those found in Brazil. Brazilian towns followed no regular plan but settlements grew up as clusters of dwellings around the major public buildings or a fort; defensive considerations were more important in the siting of towns in Brazil than Spanish America, primarily because of the threat from the Dutch and French to the north. The lack of a formal plan also reflected the less pervasive control of the Portuguese Crown over colonial affairs and the smaller emphasis it placed on urban life and on the town as an instrument of colonization.

The colonial economy

The early sixteenth century
Following the establishment of towns, *encomiendas* (Box 2.2) and land grants were allocated to those who had taken part in conquest.

Although land grants were made at the time of conquest, agriculture was slow to develop given that wealth could be acquired more easily by plundering native treasures and extracting alluvial gold. Agriculture required an investment of money and time, neither of which conquistadores could afford if they were to return to Spain with improved economic and social positions within their short lifetimes. The early sixteenth century can therefore be characterized as one of plunder, where Indian goods were expropriated and native labour was often worked to extinction. Once these immediate sources of wealth had been effectively exhausted, the Spanish sought alternative sources of profit, which they found in the development of silver mining and agriculture.

Box 2.2 *Encomienda*

An *encomienda* was a grant of Indians to an individual, who in return for providing the Indians with Christian instruction, could levy tribute from them in the form of money or goods, and until 1549 could also exact labour services. In Spain the *encomienda* had bestowed rights over land as well, but since this had encouraged the emergence of estates whose powerful feudal lords posed a threat to the Crown, in the New World the grant of an *encomienda* did not carry with it the land on which the Indians lived. Nevertheless, the allocation of *encomiendas* met colonists' demands for immediate wealth and in part satisfied their desire for the type of overlordship that characterized feudal estates in Spain.

Mining

Once the Indian treasures had been plundered and the rich and easily worked alluvial gold deposits had been exhausted, the Spanish were forced to turn their attention to the mining of mineral ores. These ores were to be found in the Andes and Mexican highlands. Whereas alluvial gold mining had required little capital outlay and had often depended on Indian slave labour, the mining of vein ores required considerable capital investment. Capital was needed for the construction and maintenance of mining galleries and ventilation shafts, and for the purchase of equipment and materials to mine and process the ores, as well as for the wages and food of mine workers. The more extensive nature of vein ores and the greater capital outlay encouraged the emergence of permanent mining camps, some of which developed into prosperous towns.

Mining, like conquest, was a joint enterprise between the Crown and private individuals. With conquest the Crown assumed ownership of all of the sub-soil, but it permitted individuals to mine ores in return for one-fifth or a *quinto* of the gold or silver produced. In order to control mining activities, all claims to mines had to be registered with the authorities, and miners swore to bring any bullion they produced to the assay office for taxing. Initially silver and gold was smelted in crude furnaces, but only rich ores could be processed by this means. In 1554 the amalgamation process was discovered which enabled poorer ores to be refined. This process required the application of mercury and salt among other things, and from the mid-sixteenth century most ores mined in Spanish America were refined by this method. Initially the mercury was imported from Almadén in Spain, but after the Huancavelica mercury mine was discovered in what is now central Peru in 1574, most of the mercury came from there. Since most ores could not be worked without the application of mercury, by controlling its sale and making the Huancavelica mine a royal monopoly, the Crown attempted to keep an account of the amount of silver and gold that was produced and ensure that it received its due fifth.

Mineral production in the sixteenth century was dominated by the silver mines of Potosí in Upper Peru and Zacatecas and Guanajuato in northern Mexico. From 1531 to 1600 over 15 million kilograms of silver were exported to Spain and in the last quarter of the sixteenth century silver bullion accounted for about 90 per cent of Latin American exports by value, with over three-fifths of the silver coming from the Peruvian mines. During the seventeenth century mineral production fell, with the decline being more marked in Peru, hence when production revived in the eighteenth century three-fifths of the silver was being produced in Mexico, and in 1800 it minted five times as much silver as had been produced in Peru in the early seventeenth century.

The location of mining activities (Fig. 2.6) was fixed by the distribution of minerals, and operations depended less on local supplies of labour, shortages of which could be overcome in many cases by importing labour from other regions or from Africa. Where possible mine owners sought forced Indian labour since this was the cheapest form of labour available. The Potosí mines were located in a region of relatively dense native population and the Spanish modified the *mita*, a system of forced labour instituted by the Inca, to supply the mines with labour. Under the *mita* each Indian village was required to supply one-seventh of its adult male population to work in the mines for a year; at its height the *mita* supplied the Potosí mines with 13,500 forced labourers. But not all mines were located in areas of dense native population. Mining generated sufficient profits, however, to enable miners to pay wages that could attract free workers, as occurred in northern Mexico, or to purchase black slaves, as was the case in the Brazilian and Colombian gold mines. In these areas, therefore, mining not only extended European colonization into remote and sparsely settled regions, but it changed the racial composition of the population.

Mining activities had a number of important political and economic consequences:

1. The profits from mining stimulated the price revolution in Europe and enabled the Spanish Crown to secure loans which were partly used to finance costly foreign wars. From the early seventeenth century Crown revenue increasingly failed to cover expenditure, but its income from the taxes on precious metals effectively enabled it to maintain its empire despite its increasing bankruptcy. To its American colonies the *quintos* represented a permanent loss of

Fig. 2.6 Colonial mining areas and trade routes

    wealth, but mining often had the effect of stimulating local economies. The mining industry required mules for transport, hides for a variety of types of equipment, tallow for candles, as well as food and clothing for mine workers.

2.   As a result mining stimulated agricultural development in surrounding or nearby regions, most notably in northern Mexico, in the Bajío of Guanajuato and the valley of Guadalajara, and in north-west Argentina, where annual livestock fairs were held from which the mines of Potosí were supplied.

3.   Mining also underlay the prosperity of many towns. Their wealth
     was reflected in their elaborate architecture and developments in
     the arts, while affluent miners and merchants formed a flourishing
     market for luxury goods, including European manufactures and
     locally produced wine and brandy. The movement of minerals and
     goods required in the mining industry, as well as luxury items,
     formed the basis of regional trade. Nevertheless, even the richest
     mining areas eventually declined, in some cases leaving ghost
     towns whose elaborate public buildings remain the only testimony
     to their past prosperity.

Agriculture

During the first decades after conquest, the Spanish and Portuguese
depended primarily for their own subsistence on crops produced by the
Indians, which they obtained as items of tribute or by trade. While more
immediate sources of wealth existed, little incentive existed for colonists
to become involved in commercial agriculture. Once these sources were
exhausted, the high prices for tropical products in Europe and the
development of domestic markets for agricultural products in the towns
and mining areas that could no longer be supplied by declining native
populations all acted as stimuli to agricultural development.

   A prerequisite for the development of agriculture was the acquisition
of land. Land grants, known as *mercedes* in Spanish America and
sesmarias in Brazil, were allocated to private individuals as soon as areas
were conquered, in the same way that lands had been distributed
following the Reconquest in the Iberian Peninsula. Most of the grants
were large, with many exceeding several thousand hectares, and in
many cases the earliest colonists appropriated all of the land within a
town's jurisdiction, leaving little for those who came later. In this way
the Spanish and Portuguese attempted to control vast stretches of
territory with a small number of settlers and at the same time meet the
demands of conquistadors for social prestige which could be acquired
through landholding. Nevertheless, at the outset land grants seldom
formed a consolidated unit, especially in areas of dense native
population, for Indian rights to the land they possessed at the time of
conquest were recognized in law. Over time, however, many landholdings were consolidated and expanded piecemeal, sometimes with
the aim of monopolizing land and driving out competition. Native
lands were often usurped, but in other cases Indians sold their lands in
order to discharge tribute debts or to meet other official or illegal
demands from royal officials or priests. Particularly active in the
acquisition of land were the missionary orders, notably the Jesuits and
Dominicans, whose estates increased in size as those owned by secular
landlords became fragmented due to inheritance laws which demanded
their subdivision amongst heirs.

   Initially agricultural labour was provided by Indians working in the
personal service of *encomenderos*, but this type of labour was banned in
1549. Subsequently, in regions of dense native population the

repartimiento, a system of forced labour similar to that of the *mita* used in mining, provided labour for agricultural activities, and it continued throughout the colonial period in Oaxaca, the highlands of Guatemala and Ecuador. Where labour was short, landowners attempted to recruit free workers by offering them incentives such as better wages, credit or plots of land on their estates. Indeed, in order to secure labour landowners often attempted to tie workers to estates by encouraging them to incur debts, a system known as debt peonage. The extent to which Indians were tied to estates by debts or other means during the colonial period has generally been exaggerated. Probably most remained on estates because they had no alternative means of surviving, and because the estates provided a degree of security, particularly in times of crisis.

In areas where Indian labour was short, but good profits could be made through the raising of commercial crops, particularly sugar, the importation of black slaves was an economic proposition. Heavy losses on the voyage from Africa, averaging over 15 per cent during the colonial period, and low reproduction rates in the New World resulted in high prices for slaves. The main areas into which slaves were imported in large numbers were the Greater Antilles, where the native population had been greatly reduced in the early years of conquest, Brazil and the Guianas. All these areas were nearer the sources of slaves in Africa and nearer European markets; although other areas, such as coastal Peru and Ecuador, were equally suited to the cultivation of tropical crops, high transport costs reduced the profitability of their products, such that there the importation of black slaves on a large scale was not a viable proposition. It has been estimated that up until 1870 about 8 million black slaves from the Guinea coast, Angola and the Congo entered Latin America, of which 5 million went to Brazil (Fig. 2.7).

The nature and distribution of commercial agricultural production was to a large extent determined by the demand for particular products and by the physical ability of areas to produce them. Many of the crops and most of the animals that were raised were not indigenous to the New World. Cacao, tobacco and dyestuffs, such as indigo and cochineal, had been produced by the Indians in pre-Columbian times. New crops – sugar, the citrus fruits, wheat and barley – were introduced from Europe in the sixteenth century. Coffee and Asian spices, such as cinnamon, arrived in the eighteenth century, introduced from the East Indies by the Dutch. Many became important export crops, but they were only selectively adopted by the Indians. Since few domesticated animals had existed in the pre-Columbian period, cattle, sheep, horses, pigs and chickens contributed greatly to agricultural production both for export and subsistence, with cattle in particular multiplying rapidly on the open grasslands.

The demand for agricultural products in Europe consisted mainly of tropical crops such as sugar, cacao and dyestuffs, as well as hides, whilst the expanding urban populations in Latin America emerged as important domestic markets for food, particularly wheat and maize, and

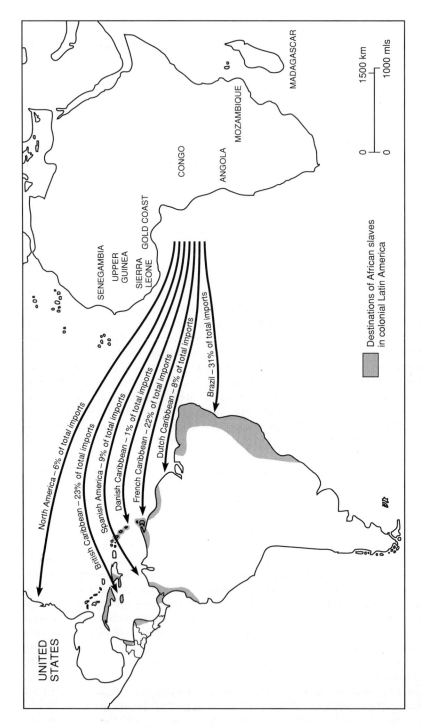

Fig. 2.7 The African slave trade to the Americas (*Source:* based on Collier, Skidmore and Blakemore 1992, p. 139)

the mining industries required hides, mules and tallow. The lack of demand for temperate products at this time meant that agricultural production in Uruguay, Argentina and Chile was essentially oriented towards the small domestic market. Demand and the physical ability of an area to produce particular crops, however, were not the only factors involved. The production of certain products, such as wool, oil and wine, for which there was a demand in Europe, was specially discouraged in Latin America because production there would have competed with that in Europe.

High transport costs generally meant that agricultural products from mainland South America could not compete in Europe with those from the Caribbean, so that agricultural production there was geared more to domestic markets in the major towns, ports and mining areas. The major market in Peru was the mining centre of Potosí, which consumed food, wine and brandy produced in the coastal oases and central Chile, and was supplied with livestock and animal products from north-west Argentina. In other areas, estates produced wheat, maize and livestock to support the towns, ports and lesser mining centres; the high Andean basins of Peru supplied the mines of Cerro de Pasco and Huancavelica, as well as Lima, whilst the highland basins around Quito and Bogotá supplied those cities and the local mining industries of Zaruma and the Upper Cauca valley respectively. The Mexican economy was more diversified. Mexico exported products to Europe and supplied the domestic market in the towns and mining areas. The commercial production of food first developed in central Mexico as population decline rendered native supplies unable to meet the growing demand in towns of central Mexico. Beginning in the Valley of Mexico, agricultural production spread rapidly into the Bajío and the Valley of Guadalajara as silver mines were opened up to the north. In the Caribbean and Brazil, agriculture was dominated by the production of sugar, although in Brazil livestock-raising developed in the interior; at first it provided mules for the sugar industry, but later, from the early eighteenth century, it supplied the needs of gold miners in Minas Gerais.

Manufacturing industry

Despite the fact that in pre-Columbian times many native societies contained skilled artisans, such as potters, weavers and silversmiths, during the colonial period manufacturing industry was slow to develop. The Spanish Crown aimed at protecting its home industries by establishing a monopoly of trade in European commodities. However, its home industries were unable to meet the demand for such items as fine cloths, leather goods, glass, pottery, weapons and general hardware, so that many goods had to be imported from other parts of Europe before being shipped across the Atlantic from Spain. Even so, the supply failed to meet the needs of colonists, who were often forced to turn to contraband trade with the English, French and Dutch in order to obtain the items they required. Although local manufacturing industries were often discouraged, and in some cases suppressed, interference in colonial industries was never as systematic or as complete as it was with trade. A large number of small-scale enterprises did develop to supply the local market, which included many Indians and persons of mixed race, who were generally unable to purchase expensive European goods. Apart from the processing of foods and beverages, and the refining of silver, the most important items manufactured in the New World were textiles and clothing. Coarse woollen and cotton cloth were produced in east-central Mexico supplying the mining areas to the north, but the most renowned area of production was in the highlands of Ecuador around Quito, Latacunga and Riobamba, from where textiles were transported to Peru, especially to the mining areas, where they competed with those produced locally.

Trade and transport

In the colonial period trade and transport revolved around export production. Apart from trade across the Atlantic, much of domestic trade was also linked to the export-economy in that it supported mining and agricultural enterprises whose production was geared to the European market. The emergence of new economic centres and the outward orientation of production altered many of the road systems that had existed in pre-Columbian times. For example, the Spanish developed the west–east Inca roads that focused on the ports, at the expense of the north–south routes that the Inca had constructed to integrate their empire. The basis of the colonial transport network was the *camino real* or royal highway; with the exception of the Magdalena river in Colombia, the Spanish did not develop river transport. Where the terrain permitted, goods were often transported by mules and ox-drawn carts, but in many areas native porters remained the basis of the transport system.

Although the new forms of transport represented an improvement over the llama – the only form of non-human transport in pre-Columbian times, which was restricted to the Andean area – mules and carts were slow and not suited to bulk transport. Mules only averaged about 19–24 kilometres (12–15 miles) a day and could carry only about 136 kilograms (300 lb). This form of transportation restricted the production of bulky and perishable goods. It meant that the commercial production of crops for the European market was not a viable proposition in remote parts of South America; even in more accessible regions, such as Mexico, agricultural production favoured those crops of high value and little weight, such as indigo and cochineal. The limitations of the transport system also effectively restricted large-scale production of crops for a mass domestic market, so that estates there remained relatively small producers whose products were rarely traded beyond neighbouring provinces.

Pursuing mercantilist policies (Box 2.3) which were dominant in Europe from the fifteenth century, the Spanish Crown established a monopolistic and protectionistic trading system aimed at securing the greatest profits for itself and for the few merchants who it entrusted

---

Box 2.3 Mercantilism

Mercantilism was the guiding principle behind the economic policies pursued by European states from the fifteenth to the eighteenth centuries. It regarded the acquisition of precious metals as a means of achieving a favourable balance of trade that would enhance state power. 'The more silver and gold, the stronger the state.' On this basis states justified their monopoly control of trading systems as well as the exclusion of foreign merchants.

---

with its operation; foreigners were not allowed to trade in Spanish America, except that they could introduce black slaves. It was envisaged that Spain would supply the colonies with manufactures, including luxury goods, in return for the minerals and tropical agricultural products they produced. European goods were to be transported on Spanish vessels manned by Spanish sailors, after having been registered in Seville, where the Board of Trade, the *Casa de Contratación* (established in 1503) and the *Consulado*, the merchant guild which arranged for their shipment, were both located. All goods destined for Spanish America from anywhere in Europe had to pass through Seville (and from 1717 through Cádiz), and as such the merchants in Seville controlled all trade with Spanish America and by setting the prices for goods, ensured maximum profits for themselves.

To safeguard goods in transit, from 1552 ships were required to sail in convoy under the protection of armed vessels known as *armadas*. Two fleets were dispatched from Spain every year. One, the *Flota*, sailed to Veracruz in Mexico, convoying ships for Central America. The other fleet, the *Galeones*, sailed to Portobello and Cartagena. From Portobello goods passed across the Panamanian isthmus to the Andean and river Plate countries. Until 1778 no trading could take place through Buenos Aires, which remained a closed port in order to prevent contraband trading with the Portuguese and northern Europeans. From Cartagena goods passed to Colombia and Venezuela. At these three ports merchants, mainly from Lima and Mexico City, monopolized the items coming from Europe and from thence transported them to the colonies selling them at highly inflated prices. The two fleets wintered in Latin America, picking up minerals and agricultural produce to transport to Spain, after which they rendezvoused in Havana before sailing across the Atlantic together. Just as the trade across the Atlantic was restricted to certain ports and controlled by a handful of merchants, so trade with the Philippines was only permitted at Acapulco.

The fleet system operated until the beginning of the eighteenth century, but it reached its peak about 1590, when the fleets comprised 30–90 vessels, after which it gradually collapsed. The fleet system was costly to maintain and the Crown's income from trade declined. Spain could not provide the manufactured goods demanded in the colonies – by the end of the colonial period 90 per cent of those exported to Latin America came from other countries in Europe.

Portuguese commercial interests focused on the East, so that the commercial system developed for Brazil was never as clearly formulated or as extensive as that developed for Spanish America. A fleet system was introduced in the middle of the seventeenth century in an effort to counteract the presence of foreigners, but it was never as restrictive as that developed by Spain, and it underwent many changes before it was finally abolished in 1765.

The peripheries

So far we have referred to the core areas where the Spanish and Portuguese established their towns and economic enterprises and drew into their orbit the indigenous population. However, many areas remained outside effective administration, since they contained only sparse native populations and did not possess resources which were currently in demand in Europe. Three types of activities occurred in these regions: first, the sparsely inhabited grasslands, particularly of Argentina, and southern Brazil and southern Venezuela, were subject to invasion by feral cattle. During the colonial period these cattle were hunted by small numbers of marginal people often of mixed race, but in the nineteenth century such areas emerged as the major ranching regions of Latin America; second, in many areas missions were founded to convert and 'civilize' the native population and form buffer zones against foreign encroachment; third, northern Europeans occupied territories where Spanish and Portuguese control was weak. The colonization of the latter two areas will be considered in more detail.

The missions

The initial conversion and 'civilization' of Indians living in remote parts of Latin America was undertaken by the missionary orders. The most important orders that established missions in Latin America were the Jesuits, Franciscans, Dominicans and Augustinians. The general strategy was that missionaries would move into an area and settle Indians in missions where they would be 'civilized' and converted to the Catholic faith. It was envisaged that after ten years the missions would be handed over to the secular authorities and the Indians would become fully fledged tribute-paying citizens. The missionaries would then move on to 'civilize' and convert Indians in more remote areas, and by so doing gradually push back the frontier of Spanish settlement. However, because of the shortage of parish priests and secular officials to administer the mission settlements, the missionaries very often remained for longer periods. Nevertheless, the missions acted as effective instruments of colonization, often being supported financially by the Crown, particularly in areas that were under threat of foreign domination, such as northern Mexico, eastern Central America, the lowlands to the east of the Andes and the La Plata region.

The task of establishing missions was not easy since the missionaries often worked with nomadic or semi-nomadic groups who were

Plate 2.1 Eighteenth-century church at former Jesuit mission of Concepción, eastern Bolivia. The missions were intended to protect the Indians from marauding slaves. The churches demonstrate indigenous craft skills harnessed to serve the new religion. (Photo Tony Morrison, South American Pictures)

scattered over wide areas and were reluctant to settle and remain in the missions. Soldiers were often required both to bring Indians into the missions and to prevent them from fleeing. Apart from converting the Indians, the missionaries sought to 'civilize' them by requiring them to live in settled villages and to practise agriculture. They suppressed polygamous practices and by establishing Indian *cabildos*, they trained them in new forms of political government. Thus, missionization resulted in native cultures being profoundly modified, if not destroyed. This, coupled with the introduction of Old World diseases, resulted in the decline and disappearance of many Indian groups. Even in Paraguay, where the Indians fled in large numbers into the missions to escape the enslaving expeditions of the Paulistas (from São Paulo), the expulsion of the Jesuits in 1767 resulted in the disintegration of the mission settlements, as individuals drifted away in search of employment on local estates or in the towns. Hence, although in the short term and in some places the missionaries may have protected Indians from exploitation, the Old World diseases they introduced greatly reduced native populations, and at best their activities only delayed the process of acculturation and assimilation.

The northern Europeans

For both ideological and practical reasons the Iberian powers sought to prohibit the settlement of foreigners in the New World. This policy was consistent with the prevailing concepts of absolute monarchy and mercantilism, and it was justified by the Papal Bulls of 1493 and the Treaty of Tordesillas of 1494 which upheld Spanish and Portuguese claims to the New World, specifically for the propagation of the faith. The northern European nations, notably England, France and the Netherlands, on the other hand, were anxious to undermine Spanish and Portuguese power in Europe, and their American colonies were the most vulnerable parts of their empires. Initially foreign activities comprised attacks on the treasure fleets and the ports, but later from the early seventeenth century interest focused on acquiring territorial footholds on the Caribbean islands and fringing mainland from which they attempted to undermine Iberian power by contraband trade and later colonization (Fig. 2.8). These areas, together with the Guianas, were those where Spanish and Portuguese control was the weakest. The English first settled in Barbados in the 1620s, but probably the most important English possession in the Caribbean was Jamaica, which was seized in 1655. Meanwhile the French had occupied Guadeloupe and Martinique and had begun to colonize western Hispaniola, whilst the Dutch had seized Curaçao in 1634. In addition, all three nations jostled to establish colonies on the weakly defended and uninviting Guiana coast. Although most of the seizures and occupations of northern Europeans in the New World were illegal, their rightful possession of these territories was often confirmed at peace treaties concluding European wars. The acquired colonies were used as bases for contraband trade with the Spanish colonies and Brazil, whilst the development of sugar and tobacco plantations using imported black slaves on the Caribbean islands and in the Guianas constituted valuable sources of income that enhanced the economic power of the northern European nations. Even where these nations did not formally acquire territorial possession, their influence was often strong. Such was the case along the Caribbean coast of Central America where the English established wood-cutting enterprises and exploited turtleshell.

Native societies under colonial rule

Iberian conquest and colonization were disastrous for the Indian population of Latin America. At the end of the colonial period even the most fortunate Indian groups who had come into contact with Europeans were less than half the size they had been at the time of the Conquest. During the first century following European contact many native societies lost 90 per cent of their populations. In Peru the population fell from about 9 million to 670,000 in 1620 and in central Mexico from between 10 and 25 million to about 1 million.

Meanwhile in the Greater Antilles and on the nearby mainland of the Caribbean most groups had become virtually extinct by the mid-sixteenth century. In the highlands of Mesoamerica and the Andes

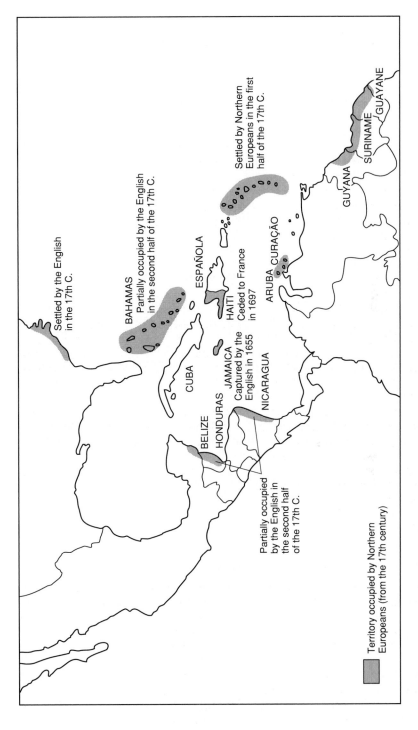

Fig. 2.8 Territory occupied by northern Europeans from the 17th century (*Source:* based on McAlister 1984)

native societies achieved a degree of demographic recovery during the colonial period, but elsewhere the decline continued.

Several factors were responsible for the decline in the native population, including disease, systematic killing, ill treatment and overwork of the Indians; racial mixing and the indirect effects of changes to native economies, societies and beliefs brought about by conquest and colonization. It is now generally recognized that the introduction of Old World diseases to which the Indians had acquired no immunity was a major factor in the decline of the native population (Fig. 2.9). The most notable killers were smallpox, measles, typhus, plague, influenza, yellow fever and malaria. There are numerous accounts of the populations of villages and whole areas being reduced by one-third or one-half as a result of epidemics, particularly of smallpox and measles, and in the early colonial period communities were devastated not once, but many times at almost regular ten-year intervals. The same demographic collapse did not occur in Asia and Africa where native peoples had generally acquired some immunity to Old World diseases which had been introduced at an earlier date through trading contacts.

Sixteenth-century observers blamed the extraordinarily rapid decline of the native population on the ill treatment and overwork of the Indians by conquistadors and colonists. There is no doubt that the Black Legend was a reality in the Caribbean where it contributed significantly to the extinction of formerly dense native populations. In 1542, as a result of representations to the Spanish Crown, particularly by the Dominicans, the New Laws were introduced which aimed at improving Indian–European relations and protecting the Indians from exploitation. Although the New Laws were often infringed, by banning Indian slavery, moderating the use of Indian labour and regulating the amount of tribute that could be levied, they did lead to a general improvement in the treatment of the Indians to the extent that the colonization of the mainland which occurred mainly after their introduction, did not result in a repeat of the demographic disaster that had occurred in the Caribbean and to a lesser extent in Middle America. Even though Indian slavery was banned in 1542, it continued in remote parts of the Spanish empire, notably in northern Mexico, southern Chile and Argentina, where hostile Indians remained a constant threat to European settlements. In Brazil Indian slavery persisted throughout the colonial period. Acute shortages of labour encouraged slave-raiding expeditions into the interior, the most notorious of which were those conducted by the Paulistas (people from the region of São Paulo). Slavery not only broke up and destroyed native communities, but led to the rapid acculturation and racial assimilation of those enslaved.

During the colonial period *mestizaje* or racial mixing had an increasing influence on demographic trends, both encouraging native population decline and retarding its recovery. The predominance of males amongst colonists to Latin America in the early years of conquest encouraged racial mixing and the emergence of *mestizos* (see Chapter 6). Even though laws discouraged contact between the races, it was

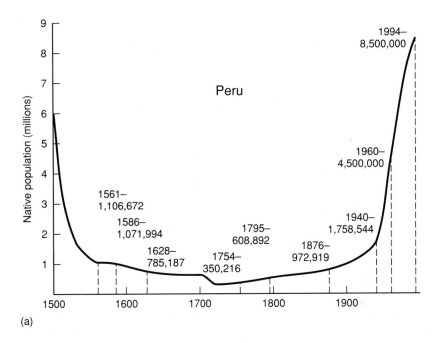

(a)

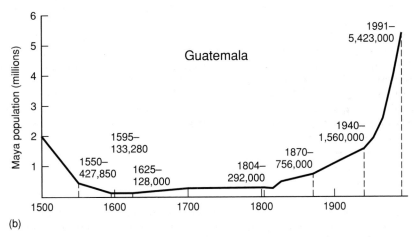

(b)

Fig. 2.9 Native population decline: (a) Peru; (b) Mayan population of
Guatemala (*Source*: (a) based on Cook 1965; (b) based on Lovell and
Lutz 1994)

inevitable given the disdain of Europeans for manual work and their
reliance on Indian and black labour. As such the major centres of
employment – the haciendas, mines and towns – emerged as racial
melting pots.

37

Equally important in contributing to the decline in the native population, but often understressed in the literature, were the indirect effects of changes to native economies, societies and beliefs brought about by conquest and colonial rule. During the Conquest Indian lands were overrun and pillaged; others were usurped by Europeans or ravaged by straying livestock. Not only was the availability of land reduced, but also, and probably more important, there were fewer people to work it. Native depopulation and European demands for tribute and labour undermined subsistence production and made it difficult for native societies to maintain terraces, raised fields and irrigation systems. While food shortages and even famines sometimes resulted, in other cases diets may have improved, most notably through the greater availability of meat, resulting from the introduction of Old World livestock, particularly chickens and to a lesser extent cattle. Native food supplies must have varied widely according to ecological conditions and European demands for land and labour. These demands were greatest in areas of commercial crop production, and often lower in areas of livestock-raising which generated smaller demands for labour and often occupied grasslands which not had been intensively utilized in pre-Columbian times. Meanwhile, in many areas, such as Yucatán, native depopulation encouraged forest regeneration. Indeed, it has been argued that the human imprint on the landscape was less visible in 1750 than in 1492 (Denevan 1992a).

Depopulation and colonial rule brought major changes to native social organization. The authority and status of native leaders was generally undermined by colonial rule, though some enhanced their positions through marriage alliances or by manipulating the colonial administrative system to their advantage. With the exception of Indian leaders, the Spanish did not recognize native social classes but rather subjected all to common laws and systems of tribute payment. Meanwhile the kinship ties of native communities were modified, and generally weakened, by depopulation, migration and resettlement schemes. In the sixteenth century depopulation prompted resettlement schemes aimed at facilitating the secular and religious administration of reduced native communities. In the 1570s massive resettlements took place in Peru that involved the movement of about 1.5 million Indians, and similar programmes were conducted in Mexico at the turn of the sixteenth century. The new settlements often contained remnants of communities with few common interests. Further social dislocations occurred as Indians fled to avoid tribute payment and forced labour, and as they were increasingly drawn to the towns, mining areas or haciendas in search of wage labour.

The psychological impact of conquest and colonization should not be underestimated, even though it is difficult to quantify. Defeated in conquest and ravaged by disease, the Indians believed that they were being punished by their gods. This, in addition to the real hardships of existence, fostered a lack of will to survive, such that in the early colonial period abortions and infanticides increased and families became smaller.

Given the level of depopulation and cultural change in the colonial period, it is remarkable that so much indigenous culture has survived. Indeed Indians still make up the majority of the population in Bolivia and Guatemala, and form a significant minority in Peru, Ecuador and Mexico. Native peoples were not passive victims of colonial rule, neither were their cultures totally destroyed. While many changes were forced upon them, they often manipulated them to their advantage, peacefully resisted by preserving their traditions in clandestine ways, and even revolted. Some Indian leaders forged early alliances with Spaniards, while native communities and individuals rapidly learned to use the Spanish legal system to defend their rights to land and to seek redress for wrongs. They selectively incorporated aspects of European culture that could enhance their livelihoods, such as metal tools, chickens, pigs and sometimes cattle, while indigenous religious beliefs and deities were often fused with Christian ones, producing a syncretic religion that was widely tolerated by the Catholic Church. At the end of the colonial period resistance exploded into open revolt. In the 1780s major rebellions occurred in Peru and Bolivia with the rebels seeking social justice and drawing inspiration from their Inca past. Other revolts occurred in Mexico, while on the fringes of the Spanish empire other groups, such as the Araucanians in southern Chile and the Jívaro of Ecuador and Peru, remained unsubdued throughout the colonial period.

Conclusion

The concentration of population and economic activities in the highlands of Middle America and the Andes in pre-Columbian times was to a certain degree emphasized during the colonial period, although not without a dramatic decline in the Indian population and the restructuring of native societies. Areas which had been sparsely settled in pre-Columbian times remained as such throughout the colonial period, with the major exception of areas where precious minerals were found. An important spatial change, however, was the development of coastal settlements which acted as political and economic control centres for the administration and economic exploitation of the sub-continent. The colonial experience for native peoples was therefore varied, reflecting the policies and priorities of the Iberian powers as well as local cultural and environmental conditions. The distribution of population and economic activities was to experience perhaps more sweeping changes during the nineteenth century, when the temperate regions, which had remained underdeveloped during the pre-Columbian and colonial periods, received an influx of capital, technology and immigrants aimed at developing the production of temperate agricultural products for which a demand had developed in Europe with the Industrial Revolution.

Further reading

Bethell L (ed.) (1985) *The Cambridge History of Latin America*, Volumes 1 and 2. Cambridge, Cambridge University Press. (Essential reading on the pre-Columbian and colonial periods. Collections of excellent and up-to-date essays by the most eminent scholars in their fields, selections of which have been published in paperback as *Colonial Spanish America* and *Colonial Brazil*. Particularly useful are the chapters on colonial mining, urban development, the hacienda and Indians under Spanish rule.)

Bolton H E (1917) 'The mission as a frontier institution in the Spanish American colonies', *American Historical Review* **23**: 42–61. (The classic paper on the subject.)

Butzer K (ed.) (1992) 'The Americas before and after 1492: current geographical research', *Annals of the Association of American Geographers* **82** (3). (Special edition to mark the Columbus quincentenary.)

Denevan W M (1992a) 'The pristine myth: the landscape of the Americas in 1492', *Annals of the Association of American Geographers* **82** (3), 369–85.

Denevan W M (ed.) (1992b) *The Native Population of the Americas in 1492*, 2nd ed. University of Wisconsin, Madison. (A reprint of the first edition, with a new introductory section containing revised estimates.)

Dobyns H F (1966) 'Estimating Aboriginal population', *Current Anthropology* **7**, 395–449.

Fiedel S J (1992) *Prehistory of the Americas*, 2nd ed. Cambridge, Cambridge University Press. (A broad-ranging volume with particularly useful sections on the origins of agriculture and the emergence of native American civilizations.)

Gibson C (1966) *Spain in America*. New York, Harper & Row. (An excellent and very readable introduction to colonial Spanish America.)

Hennessey A (1978) *The Frontier in Latin American History*. London, Arnold.

Kicza J E (ed.) (1993) *The Indian in Latin American History: Resistance, Resilience and Acculturation*. Wilmington, DE, Scholarly Resources. (A collection of reprinted articles covering the colonial period to the present day.)

Lockhart J and **Schwartz S B** (1983) *Early Latin America*. Cambridge, Cambridge University Press. (A broad introduction to the colonial period, which is particularly useful for colonial Brazil.)

Rout L B (1976) *The African Experience in Spanish America, 1502 to the Present Day*. Cambridge, Cambridge University Press.

Steward J H and **Faron L C** (1959) *Native Peoples of South America*. New York, London and Toronto, McGraw-Hill. (Although over 30 years old, it remains a useful comprehensive introduction to native cultures in South America, past and present.)

CHAPTER 3

The rise of industry in a world periphery

John Dickenson

The early economic development of Latin America is generally interpreted as part of a process in which the continent was subservient to
the interests of a capitalist core in Europe and North America. Though
the form of such dependence varied, from a colonial linkage *c.*
1500–1810, via mercantile empires from 1810 to 1890, to that of
peripheral capitalism, Latin America was categorized as part of a periphery or semi-periphery to the world core region of advanced
capitalism, in which the dynamism for economic change lay outside,
rather than within the continent. In an international division of labour,
the periphery functioned as a supplier of primary products for the core,
and as a market for its industrial goods. During the colonial period,
economic control was exercised directly by the metropolitan power.
Later, control was indirect, via foreign investment in productive activity,
trade and infrastructure. Even where, in the era of peripheral industrial
capitalism, manufacturing developed, it was claimed to be in the
interests only of the national élite, or of multinational companies
(MNCs), to fulfil their needs for resources, cheap labour and markets.

Such interpretations can be sustained in Latin America. However,
such retrospective critiques neglect to suggest what alternative strategies
might have been followed, or what levels of development might have
been attained without such intrusions and linkages to the capitalist
system. Moreover, recent scholarship suggests that not all the
continent's industrial development was externally driven. Some pre-
colonial craft industries survived the European conquest, to which
colonial, but locally stimulated, artisan activities were added. Such
workshop industries have been described as 'proto-industrial'. They
provided a degree of small-scale but autonomous manufacturing, and
laid a foundation for the emergence of a modern industrial sector, even
in industries such as textiles, where the factory products of the core
might have been expected to be most intrusive.

This chapter outlines the rise of industry in Latin America. It
indicates the role played by the core in stimulating export-led
processing industries, the provision of infrastructure and capital

essential for industrialization and, towards the end of the period covered, the advent of large foreign companies which later evolved into MNCs. However, it also suggests that there was an element of autonomous industrial activity, albeit rudimentary, which can be traced back to the colonial period. In addition, it discusses the rise of economic nationalism in the 1930s, when a desire to create economies less dependent on primary products and on markets in the capitalist core, saw some commitment to domestic industrial development.

Pre-colonial roots

As noted in Chapter 2, at the time of the Iberian Conquest there were sophisticated Indian civilizations in certain parts of the continent. Their skills included artisanal activities such as pottery, metal-working and textile-making. The Aztecs worked silver, copper and other metals, to make ornaments, weapons, tools and farm implements, while henequen and cotton provided the basis for cottage industries producing clothing and blankets. In the Andes, the Incas had developed production of metals, textiles, clothing and pottery. Much of this activity was small scale, though there is evidence that the Incas created large textile and pottery workshops on the shore of Lake Titicaca. Such activities were dependent on local raw materials, used simple technology and served local markets. However, their concentration in areas of dense Indian settlement prefigured the location of colonial manufacturing, and consumption goods, produced in small units of production, were to be characteristic of post-Conquest industrial activities.

Colonial beginnings

The nature of Iberian rule was significant in shaping the pattern of industrial activity in Latin America. The earliest industries were limited to the processing of primary products from agriculture and mining prior to export, and to making goods necessary to sustain the local population, namely food, clothing and shelter. In consequence industry developed in a limited range of locations – at the point of production of export commodities, such as the sugar zone of north-east Brazil and the mines of Mexico, Upper Peru and Minas Gerais; at the ports of Veracruz, Portobello and Salvador; or in the cities of highland Mexico and the central Andes.

The large landholdings which dominated colonial agriculture were often almost self-sufficient for their industrial needs. Besides processing export and subsistence crops, they employed carpenters and smiths to make farm implements and maintain their mills, made tallow and soap, and had small tanneries and textile plants.

The extraction of precious metals, the other prop of the colonial economy, required processing. The mining areas of Mexico, Upper Peru, Brazil, Honduras, Colombia, Ecuador and Chile formed nodes of economic activity and centres for small-scale but diverse industries, and created extensive linkages for foodstuffs. As early as 1600 there were over 400 silver refineries in Mexico and Potosí, and Brazilian gold had similar consequences, with the establishment of government gold foundries. Craft activities linked to mining and smelting, or serving the population they attracted, included metal-working, carpentry, the making of soap, wax, tiles, lime, leather goods and clothing.

Trade with Spain stimulated shipbuilding. As early as 1530 ships were being built in Nicaragua, and other shipyards were established along the coasts of Cuba, Jamaica, Puerto Rico, Mexico, Guatemala and El Salvador, at locations influenced by the availability of raw materials – timber, cloth, rope and pitch. By the mid-seventeenth century one-third of Spain's Indies fleets were being built in the Caribbean. In Brazil shipyards were soon established in Belém, Salvador and Rio de Janeiro.

Shipbuilding, crop processing, and mining and their ancillary industries were all linked to the export economy. Spain sought to protect its industrial exports to the colonies, but there were some pioneer local industries, of which the most significant was textiles. This grew out of Indian textile-making traditions, producing coarse cloth for the urban markets and mining areas. Three major centres emerged: in central Mexico, in the towns of Puebla, Mexico City and Tlaxcala; around Quito in highland Ecuador; and Córdoba and Tucumán in northwestern Argentina (Fig 3.1). Spanish introduction of sheep provided a new raw material, making woollen goods of great significance in highland areas, particularly around Puebla and Quito. More substantial units of production, *obrajes*, or textile workshops, developed. They were urban-based, using imported technology, and have been described as genuine factories, the largest of them employing hundreds of Indian or African slaves. By the early seventeenth century there were over 100 *obrajes* in Mexico and 50 round Quito, and in the eighteenth century new centres emerged such as Cajamarca, Cuzco and Querétaro. In 1790 the latter had over 3,000 textile workers in *obrajes* and small workshops.

Urban growth stimulated a range of artisan industries, including woodworking, lime, tile and brickmaking, tanning, flour milling, metalworking, and clothing and footwear. These activities were carried on in workshops or in small factories. Some commentators have suggested that this represented a form of proto-industrialization, a preparatory phase for subsequent industrialization. Certainly it can be argued that these pioneer activities helped to shape the continent's industrial landscape, with the emergence of rudimentary manufacturing in highland Mexico and the Andes, and in north-east and south-east Brazil.

Incipient industrialization in the early nineteenth century

Most of Latin America secured its political independence before 1830 but the basis for independent economic development, much less

Fig. 3.1 Textile centres in colonial Spanish America

industrialization, was limited. Instead, other European powers, and later the USA, took on major roles in exploiting the region's primary products and seeking markets for their own increasing industrial output. In consequence, the pattern of external dependency was intensified.

The new trading relationship was essentially a product of the then current notions of free trade. The evolution of a world economic system and of an international division of labour heightened Latin America's role as supplier of primary produce to the markets of Europe and later North America, and as a recipient of their manufactured goods. Such a relationship was a stimulus, as in the colonial period, to export-commodity processing activities and to necessary domestic industries; but it also provided some basis for the gradual evolution of more substantial industrialization.

The economic advance of Europe and the USA generated not only a need for raw material inputs and for markets for output, but the means by which these transactions could be carried out, with the development of steamships, and an increase in the scale and speed at which goods could be transported. Between 1840 and 1913 the world merchant fleet increased almost seven-fold. The development of the railway had similar consequences for land transportation, opening up and integrating new raw material sources and markets.

Commodity trading and transport

The ruling élites espoused these trends and initiatives, seeing the role of their countries as supplying the raw materials they could produce cheaply, in return for manufactured goods they lacked. Many saw this relationship as being the means to introduce European and North American 'progress' into Latin America and, of course, to provide themselves with wealth from the sale of sugar, coffee, minerals and other products.

The commodities sought by the neo-colonial powers took three forms:

1. Temperate agricultural products were required to meet the expanding food needs of their urbanizing populations. Demand for meat and grain prompted an extension of the frontier of exploitation and settlement, especially in Argentina and Uruguay, and was closely associated with the introduction of new modes of transport and farming techniques.
2. Secondly, demand for tropical crops, such as sugar, coffee, cacao, bananas and tobacco, created new areas of large-scale monoculture and generated new patterns of settlement and transport.
3. Basic minerals such as guano, nitrates, copper and tin provided the other resource sought by the developed world. These commodities became major elements in the trading pattern of Latin America, profoundly influencing the scale and nature of development in the nineteenth and early twentieth centuries.

A crucial element in their exploitation was the improvement in transport. In the early post-colonial period its deficiencies tended to perpetuate the pattern of concentration of economic activity close to the coast, and in the principal cities such as Rio de Janeiro, Buenos Aires,

Santiago and Lima. The introduction of the railway, however, brought profound change, facilitating the opening of interior areas and resources.

The first significant railway development was in Cuba, in 1837, where it was introduced to transport sugar and coffee. More general development began after 1850. In many areas considerable technical ingenuity was necessary to overcome such obstacles as the Andes, the Serra do Mar in Brazil and in the Sierra Madre in Mexico. The completion of railways for example the Santos–São Paulo (1868), Veracruz–Mexico City (1873) and Mollendo–Puno (1877) were of major significance for resource exploitation and development.

As Fig. 3.2 indicates, the most substantial railway systems were built in Argentina, Brazil and Mexico, which in 1940 accounted for 75 per cent of the region's network. It is also evident that, in most cases, the bulk of the railways were built between 1880 and 1920. The networks were built primarily for exploitative purposes, intended to facilitate the removal of primary produce to the ports and abroad, rather than provide co-ordinated internal transport systems. Differing patterns reflect contrasts in the nature of the export economy, with single lines providing access to mines, and more complex nets serving the more extensive production of grains, coffee, cattle and other agricultural commodities.

Though governments encouraged railway building, often with generous concessions and guaranteed profits, foreign investment and

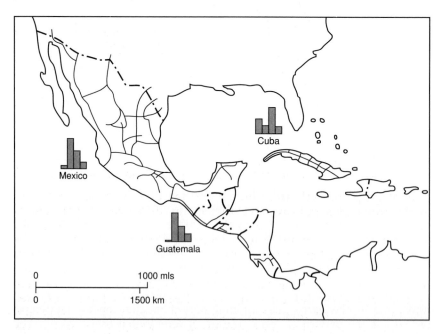

Fig. 3.2   **(a)**

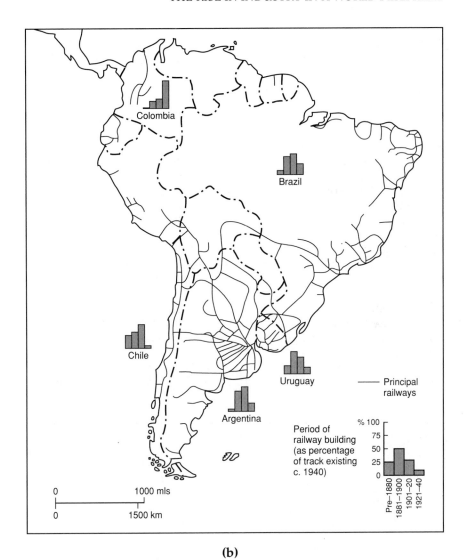

(b)

Fig. 3.2 The evolution of the railway system to 1940: (a) Central America; (b) South America (*Source*: based partly on data from Mitchell 1983)

technology, particularly from Britain and the USA, was very important in the continent's railway development. In 1914 over 40 per cent of British investment in South America was in railways, controlling, for example, the bulk of the Argentinian and Uruguayan systems. The railways confirmed and strengthened the spatial pattern of development begun in the colonial period, paralleling its routeways and increasing the significance of existing coastal settlements, although the exploitation of new resources did open up new land and extended the area integrated into the international economy. Foreign capital was also significant in improving ports and shipping, providing urban services such as transport and electricity, and in direct investment in mining and some industries. Such developments facilitated Latin America's role as a supplier of primary products, so that by 1914 it was a major source of sugar, cereals, coffee, cacao, livestock products, rubber, fertilizers and minerals such as tin and copper.

Export-based industries

The detailed spatial pattern of these activities and their contribution to incipient industrialization varied. Temperate and tropical agricultural produce initially made limited contributions to industrialization, but some were significant stimuli to factory production. A striking example of this process and pattern comes from the pastoral industries of the grasslands of Argentina, Uruguay and southern Brazil. The rudimentary extensive pastoral activity of the colonial period, with cattle being slaughtered for their hides, was replaced by more ordered cattle-raising and meat-processing, with the development of the *saladeros* (meat-salting factories). The earliest developed at Sacramento on the Uruguayan coast and along the north shore of the Plate estuary, then spreading to Argentina and Brazil (Fig. 3.3). The salting and drying of beef provided low-grade meat for the slave populations of Brazil and the Caribbean, and also yielded hides, grease, tallow, bone meal and gelatin.

Expansion of this pastoral-based industry was the product of a complex range of influences – rising demand for meat, hides and wool in industrializing Europe, improvements in livestock, the development of railways and shipping, and technical innovations including barbed wire, canning and refrigeration. Earliest changes were associated with sheep, from the 1850s, as a reflection of rising demand for wool in Europe. Wool production generated little local industrial activity, but the introduction of dual-purpose sheep provided a base for meat-processing industries.

A crucial problem for the nascent meat export trade was the transfer of the product some 11,000 kilometres to North Atlantic markets. Salt beef was not overly palatable, and technical innovations provided the basis for change. The Liebig process, producing meat extract, was developed in 1847 and resulted in the establishment of factories in Entre Ríos and Uruguay.

Tropical export crops also supported some limited processing activities. Sugar processing, for example, was carried on in relatively small-scale sugar mills (*ingenios*) in the Caribbean and north-east Brazil. In Cuba in 1860 there were some 2,000 such mills.

The agricultural export sector tended to generate a diffusion of economic activity, but the third element, of mineral exports, fostered concentration. This was inevitable, given the pattern of distribution of resources. In contrast to the colonial experience, the new mineral exports were mundane in character, providing inputs for the fields and factories of industrializing Europe and North America. Among the earliest of these was the export of guano as a fertilizer, from islands off

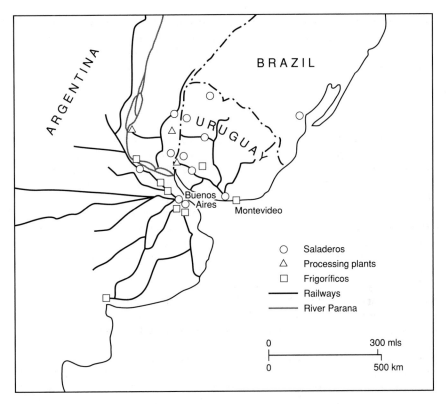

Fig. 3.3 The location of meat-processing centres in the Plate area in the period 1800–1914 (*Source*: based on information from Ruano Fournier 1936 and Hanson 1938)

the Peruvian coast. This required no processing, so that its direct impact between 1840 and 1880 in generating industry was negligible. However, output provided a large share of government revenue, and thus capital for investment in mainland agriculture, roads and railways. The boom declined with exhaustion of the resource, and competition from the nitrates of the Tarapacá and Antofagasta areas of the Atacama desert.

Autonomous development: the curious case of textiles

Though the growth of industry in the early independence period lay mainly in export processing activities, there was some autonomous development. This was most evident in the textile industry, which highlights a contradiction in interpreting the industrialization process in Latin America. As a leading industry in the Industrial Revolution of Europe it was an obvious pioneer of manufactured goods exports, and the simple nature of Latin America's textile industry made the continent a potential market for British mills. Indeed, in 1840 Latin America took

49

one-third of Britain's cotton exports; in the case of Brazil, during the period 1850–70 textiles provided 70 per cent of imports from Britain.

Yet a local textile industry, supplying local markets with the cheapest goods, survived and in some cases flourished. Thus, although imports were blamed for the demise of local industries, Cordoba continued to supply the Plate market until the middle of the century, and in Peru, though coastal markets were lost, the highland artisan industry survived. Chile retained 80,000 textile workers in 1855 and in Minas Gerais, over half the female labour force was engaged in domestic textile production. These activities were protected by interior locations, away from the ports; only as transport improved were they threatened.

There was, however, not merely the survival of traditional textiles, but the installation of a 'modern' sector. Mexico's first cotton mill began to operate in 1833 and by 1845 there were 50 mills producing cheap cotton cloth. Puebla, which had been a major centre of *obraje* textiles, had skilled workers who could adapt to an industrial technology, and an established regional market existed. New centres also emerged in lowland Veracruz, which produced raw cotton and offered water power sites. Mills were established in Brazil in the 1840s and by 1866 there were nine, five of them in Bahia, employing 800 workers.

The coming of the railway also prompted the establishment of workshops, foundries and machine shops to maintain locomotives, and in some cases to make equipment. Such activities fostered metal-working and engineering skills. In Brazil, the first railway was completed in 1858, and by 1861 there were ten foundries in Rio de Janeiro.

Urban industries

In other industries small-scale factory production began, in which industrial characteristics of concentrations of workers, specialization of labour, and use of machinery were identifiable – in furniture, hat and shoe-making, food processing, and brick and glass production. Craft shops, employing 10–20 workers, produced clothing and footwear, furniture, leather goods, soap and candles, and foodstuffs.

Such activities, serving expanding urban populations, were significant. By 1870 Mexico City, Havana, Buenos Aires and Rio de Janeiro all had over 200,000 inhabitants, Lima, Montevideo, Recife, Salvador and Santiago over 100,000. Port cities such as Guayaquil and Valparaíso had also expanded. Their basic needs were met by artisan industries producing foodstuffs, beer, cigarettes, leather goods, carpentry, and clothing and footwear.

Industrial survival and development: a paradox?

There was then, in this phase, a paradox. Standard interpretations identify this as a period in which either rudimentary activities surviving from the colonial period were destroyed by fierce competition from European (and specifically British) manufacturers or, indeed, where manufactured goods became available for the first time. Latin America provided a major market for such goods, and these markets were of great significance to expanding British producers. However, a concomitant demand for primary products, whether as raw materials for industry or as foodstuffs for a rapidly growing urban working class in Europe, stimulated export-processing industries. In addition, not only did craft textiles and some other industries survive during this period, but factory-scale production was initiated, for example in textile production in Mexico, Brazil and Peru. Even in the capital cities, where foreign competition might have been expected to be most intrusive, artisan activities persisted.

The dawn of the factory age 1870–1914

The late nineteenth and early twentieth centuries have been described as the golden age for foreign investment in Latin America. It is during this period, for many commentators, that the continent became firmly enmeshed into the world economy, through the intrusion of foreign capital, markets, transport, technology and communications. A persuasive case can be made for such an interpretation. The continent became a major market for British, European and later American manufactures, and a major source of minerals and temperate and tropical crops for the booming industries and population of those countries. It was also a leading recipient of foreign investment, in the productive sector and in the provision of transport and urban infrastructure. Such linkages, as raw material supplier, market and source of profit, and the creation of services to underpin these activities, can all be seen as drawing the continent firmly into the web of high capitalism. Such a critique presumes that no benefits accrued to Latin Americans from these experiences. Yet there patently were benefits from the creation of infrastructure and the generation of industrial output, employment and profit. Moreover, not all development was dependent on external influences; there was local initiative towards creating a manufacturing sector.

The range and volume of Latin America's export commodities rose, and foreign goods and services were increasingly significant imports. In 1913 Latin America accounted for 10 per cent of Britain's foreign trade and 20 per cent of its investment. The pattern of investment was not, however, uniform; it concentrated in certain sectors and countries. Around 1900 75 per cent of British investment was in Argentina, Brazil and Mexico, and a further 18 per cent in Chile, Cuba and Peru, mainly in government loans, railways and public utilities. Eighty per cent of American investment was in Mexico and Cuba, principally in railways, precious metals and sugar, though prior to the First World War Chilean and Peruvian minerals also proved attractive. French and German investment focused on Argentina, Brazil and Mexico. If trade and investment can be seen as exploiting Latin America, they also had local consequences. Export-led expansion required improvements in transport

Plate 3.1   The Belgrano railway station, Tucumán, north-western Argentina.
(*Source*: Photo Tony Morrison, South American Pictures)

and communications, which included shipping, railway construction, port modernization, telegraphic links and public utilities. These were major innovations, and if the capital and know-how was foreign, such projects created jobs, brought skills and provided internal opportunities.

Expansion of the railways was a crucial element, reducing the costs and improving the speed of transport. They tended to reinforce rather than diversify the continent's basic economic geography, focusing on the existent hubs of the capitals and ports, such as Veracruz, Cartagena, Guayaquil, Lima-Callao, Valparaíso, Buenos Aires, Montevideo, Santos, São Paulo and Rio de Janeiro. However, if the foci remained the same, the railways did extend the frontier, opening up Patagonia, the Llanos and northern Mexico to pastoralism, the pampas to cattle and temperate crops, and the planalto of São Paulo and lowland Middle America to tropical crops such as coffee and fruit. The search for basic minerals, and later petroleum, prompted economic development in north Mexico and north Chile, highland Peru and the Caribbean margins.

Large-scale immigration was another 'foreign' dimension which was significant during this period, but one which must be seen as having greater consequences internally than externally. This took place to settle new frontiers and, in Brazil and Cuba, also to replace slave labour. Over the period 1880–1930 there were some 3.8 million migrants to Argentina, 3.9 million to Brazil, with lesser numbers to Uruguay, Chile and Cuba.

Italians and Spaniards were the principal groups, but other sources were Germany, eastern Europe, Russia, Portugal, France and Britain.

The immigrants made considerable impact by bringing skills and creating additional markets for agriculture and industry. Immigrants were early entrepreneurs in São Paulo and Rio Grande do Sul, and in Mexico were responsible for water-powered cotton and woollen mills around Orizaba.

During this period natural increase also contributed significantly to the continent's population growth, from 30 million in 1850 to 104 million in 1930. Over that period the populations of Argentina and Uruguay increased ten-fold, those of Brazil, Chile, Colombia and Cuba, three-fold.

Export demand, improved infrastructure and population growth created new stimuli for industry. The perception that the continent's economic interests were best served by supplying raw materials in return for imported manufactures persisted, but a pro-industry lobby began to develop, some tariff protection was provided for domestic industry, and industrial entrepreneurs began to emerge from the established and the immigrant populations.

These influences served to stimulate industrial activity – in the export sector; in the creation of industries serving the domestic market; and in the workshop urban–industrial sector.

The export industries

Between 1840 and 1880 the *saladeros* increased substantially in size, with greater mechanization to process carcasses and by-products. The production of salt beef became a factory industry, integrating cattle raising, slaughter and processing. After 1870 refrigeration began to transform the industry, towards shipping chilled meat – initially mutton, because sheep carcasses were easier to handle, and then beef. By 1907 frozen beef and mutton provided 71 per cent of Argentina's meat exports, and salt beef only 4 per cent. Refrigeration factories (*frigoríficos*) began to be constructed. Argentina's first was at Buenos Aires, in 1882, followed by others along the Plate shore over the next 25 years. Uruguay's first *frigorífico*, at Montevideo, dates from 1904 and Brazil's, in Rio Grande do Sul, from 1909. This pattern reflected the movement of cattle from the interior to fattening pastures near the factories, and the need to ship the processed meat speedily to market. Wheat cultivation also began, and Argentina increased its acreage from 800,000 hectares in 1888 to 3.2 million in 1900. Initially wheat was exported as grain, but flour-milling became an increasingly important industry, for the domestic and export markets. This began as a raw-material located industry – in 1889 two-thirds of the country's steam-powered mills were in Santa Fé province, the leading wheat producer, but by 1914, 55 per cent of flour production was at Buenos Aires, the principal domestic market and export port.

Similarly, sugar processing became more complex, increasing in scale. Steam engines were introduced, and machinery rather than manual labour came to be used. Such changes required large-scale investment and fostered larger plantations and mills. Such units, the *usinas*, emerged after 1875, though it was not until the twentieth century that they became the leading source of sugar. In Cuba large mills (*centrales*) began to develop after 1900, particularly in new cane lands in the east of the island. These tended to be foreign-controlled and associated with very large landholdings, in contrast to the smaller mills and plantations in the older areas. Between 1902 and 1927 over 60 such *centrales* were built, principally with US capital, and shipped out sugar via the dense rail net and a series of specialized ports such as Nuevitas and Antilla (Fig. 3.4). There were similar processes elsewhere in the Caribbean and in Peru and Mexico.

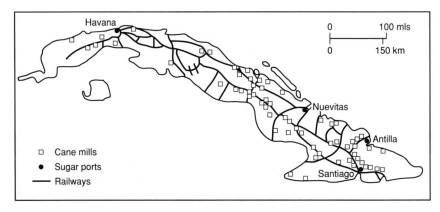

Fig. 3.4   Cuba: sugar *centrales* built after 1902 (*Source*: adapted from West and Augelli 1966: 112, 116)

Coffee and industry

The emergence of coffee as a significant export crop gave considerable significant direct and indirect stimuli to industrialization. In Colombia coffee production rose from about 6,000 tonnes in 1880 to 66,000 tonnes in 1915, and increased its contribution to export earnings from 12 per cent to over 50 per cent. Expansion was concentrated in Antioquia and formed part of a small-farm economy. Coffee processing requires only simple equipment – depulping and drying machinery, and bags for the dried beans. This encouraged simple industries producing machinery and sacking, the latter initially as an artisan industry and later in urban factories. In addition, small-farm production meant that the farmers were less prone to the 'conspicuous consumption' of imported goods characteristic of the élite. Instead their coffee income provided demand for local textiles and other necessities. This encouraged the development of craft textile activities, leather-working, pottery and iron-founding. In turn, rising prosperity and the development of a railway system encouraged factory-scale industries such as brewing, sugar refining and sweet manufacture, and textiles, all established in the first decade of the twentieth century, with Medellín as a major focus.

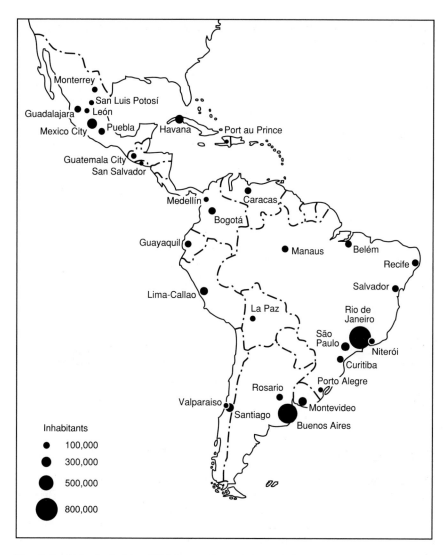

Fig. 3.5 Principal cities 1900 (more than 50,000 inhabitants)
(*Source*: based on data from Mitchell 1983: 101–8)

A very similar process, though on a larger scale, took place in south-east Brazil, especially in São Paulo state, where coffee had a major impact from the 1850s. Coffee generated capital from export earnings, land sales and wages, immigration (1.2 million immigrants between 1896 and 1906) provided a growing free market and some industrial skills, and there was major infrastructural development of railways, and of cheap hydroelectric power from 1899. These factors provided the basis for early industrialization, with metal-working and textiles for

55

coffee and other agricultural products, and food, clothing and building materials for local use.

São Paulo's railway system was also a key element in the state's industrialization, facilitating not only the rapid advance of the coffee frontier and the export economy, but also the provision of centres of metal-working and fostering of engineering skills in railway workshops. In addition railheads and junctions provided nodes of urbanization and foci for industry. However, the focusing of the rail net on the town of São Paulo (which increased in size from 65,000 to 579,000 inhabitants between 1890 and 1920) made it the principal industrial centre, with 121 mechanized factories by 1895.

Other agricultural exports included coffee from Costa Rica, Guatemala and Mexico, sugar from Peru and the Guianas, and rubber from Amazonia. The expansion of tropical and temperate agriculture brought some industrial activity to these areas, though generally of a fairly rudimentary nature. More substantial industries, whether processing raw materials for export, or imports for the domestic market, tended to concentrate in the principal ports.

Mineral-based development

In contrast, the expansion of the mining sector created new growth points, but though the location of deposits in remote and difficult areas prompted nodes of activity around mines and processing plants, there was little local spin-off. In northern Chile the exploitation of nitrates in the Atacama desert after 1883 generated a classic and ephemeral exploitative economy, in which the nitrates were mined, refined in *oficinas* and moved by rail to a series of ports. In an essentially hostile environment this was an alien activity. It remained viable so long as resources remained and there was adequate demand for the product. Foreign capital was crucial in these developments. In 1901 over 80 per cent of the investment in the industry was in foreign hands (British, German and Spanish); only 15 per cent was Chilean. At its peak during the First World War, there were some 65,000 workers in the industry and a regional population of over a quarter of a million. However, competition from synthetic nitrates destroyed Chile's monopoly and by the 1930s many of the *oficinas* had closed. Northern Chile was also the scene of small-scale copper production from the mid-nineteenth century, but after 1870 mines were mechanized and steam power installed. Modern smelters provided Chile's first large industrial enterprises. Such innovations in larger-scale mining and processing techniques involved foreign capital.

Similar features characterized other mineral-based activities – in Peru, Mexico and Brazil, for example, involving export orientation, foreign capital, spatial concentration, subservient transport systems and limited industrial spin-off.

Internal industries

There is clearly evidence of incipient industrialization in Latin America by the late nineteenth century, not only in the processing of export commodities, but in production for domestic demand. The continent had begun to benefit from the diffusion of European and North American innovations, such as railways, electricity and medical advances, though the distribution of these benefits was uneven. Some areas and segments of society remained beyond their impact, but rising urbanization and the emergence of an urban middle class generated pressure for modernization. Economic and cultural contact with Europe and North America prompted a desire for progress, including industrial development.

Pioneer industries were those with high weight-to-cost ratios able to compete with imported manufactures, and which used local raw materials or imports which gained bulk during processing. In spite of evident competition from Europe, textiles were a leading sector in this national industrialization. Mills depended on local raw materials, mostly cotton, water power sites, and the existence of markets for cheap textile goods. Mexico had been a pioneer, and by 1877 there were over 90 mills, with an average of over 120 workers; by 1910 there were over 120 mills, averaging 260 workers. The development of hydroelectric power in the 1890s provided energy for mills in Puebla, Orizaba, Jalapa and Mexico City. New sources of cotton developed in northern Mexico and the evolving rail net provided access to raw material and markets. In Brazil the number of mills rose from 9 in 1866 to 44 in 1881, in Bahia, Rio de Janeiro, São Paulo and Minas Gerais. In some cases mills were at new, water power, sites, creating company towns; elsewhere they dominated the urban neighbourhoods in which they were located.

Large units of production also developed in the brewing industry, located adjacent to the principal markets, such as Buenos Aires, São Paulo and Mexico City, and began to replace imported beer. Other industries which increased in scale, and used domestic raw materials to supply internal markets included flour-milling and shoe-making.

These industries required capital goods, which were imported, but also began to be produced within Latin America. By 1914 Argentina, Brazil and Mexico had large foundries and some iron and steel-making and engineering.

Artisan survivals

Beneath these tiers of export-led and larger-scale domestic industry, the urban-supply and artisan industries persisted. The cities and towns had a range of such activities. Small 'factories', employing less than five workers, were typical, producing foodstuffs, tobacco, leather goods, building materials, furniture, clothing and footwear, soap, candles, metal goods and other necessary items.

By 1914 Argentina, Brazil and Mexico, and to a lesser extent Chile, Colombia and Peru, had made some industrial progress. Within these

countries the spatial pattern of industrial activity was one of concentration, with the export industries at the point of production or the
ports, and industries concerned with the domestic market located at
the principal concentrations of population (Fig. 3.5). South-east Brazil,
the valley of Mexico and Buenos Aires province were the main foci of
industrial development, together with the principal cities of the other
republics. Even so, the total industrial labour force remained modest;
Argentina's second census, in 1895, recorded 175,000 workers, and
Brazil's first industrial census in 1907, 150,000. This compared with
Britain's industrial workforce in 1911 of 8 million employees.

Progress in the early twentieth century

The expansion, diversification and spatial change which took place in
the industrial sector in the early twentieth century derived from several
stimuli.

A widening resource base

An important element was the identification of a wider resource base.
Nineteenth-century economic theorists argued that cheap and abundant
power was an essential element for industrialization, yet the continent
was apparently deficient in the fundamental fuel, coal. In 1900 only
Mexico and Chile had coal industries of note, though producing only
about 350,000 tonnes each, against United States output in excess of 240
million tonnes. Before 1940 production also began in Brazil, Colombia
and Peru, and provided some basis for industries using coal as a raw
material or fuel.

Of greater significance was the utilization of alternative sources of
energy. With the recognition of petroleum as a source of fuel and power,
some countries proved to be well endowed. Mexico was the first
significant producer, from 1901. Large-scale production began in 1910,
and between 1918 and 1927 Mexico was the world's second largest
producer, after the USA, eventually losing its position in Venezuela.
Commercial production began in Argentina in 1907, Venezuela in 1917
and Colombia in 1921, and by 1940 these countries, together with Peru,
Ecuador and Bolivia were significant producers. In some cases oil
provided an important export commodity, and in others an alternative
to deficient solid fuel resources. The development of oil refineries and of
petroleum product industries provided the stimulus to new areas of
industrial activity, in the Mexican Gulf Coast, around Lake Maracaibo,
in western Ecuador, and in the Chubut and Mendoza areas of
Argentina. The industry is also significant in that it was amongst the
first in which the state began to take a direct interest in the organization
of the productive sector. Although many oilfields were initially
exploited by foreign companies which were to become the seven
'international major' oil firms (Esso, Mobil, Gulf, Chevron, Texaco, Shell
and BP), several countries were pioneers in the expropriation of these

interests, and the establishment of domestic companies to control the production of oil. This action, by Argentina and Chile in 1927, Bolivia in 1937 and Mexico in 1938, reflected an emerging view that certain sectors of the economy, such as mining, energy production, steel-making and some strategic industries, were of such crucial importance that their development must be in the hands of government and not of foreign interests.

The continent's needs were also assisted by the recognition of its considerable water power potential, estimated at 13 per cent of the world total in the early 1940s. As a source of power, hydroelectricity influenced the location of some large energy-consuming industries, but more significantly in supplying the electricity needs of factories in the major industrial centres. The structure of development before 1945 varied, with some significant involvement of foreign capital, such as the major stimulus provided to the industrial growth of São Paulo and Rio de Janeiro by the American and Foreign Power Co. Inc. and the Canadian-owned Brazilian Traction, Light and Power Co. Ltd. By the 1940s these companies were providing 80 per cent of Brazil's electricity. Elsewhere, as in Chile, Colombia and Uruguay, government was active in encouraging hydroelectric projects, while in other cases individual concerns developed sites to meet their specific needs. Thus in the late 1930s Peru's largest hydroelectric plant was owned by the Cerro de Pasco company.

The mineral endowment was also becoming more evident, with new sources of iron, manganese, copper and tin and also newly useful materials such as bauxite, nickel and ferro-alloys. Though their exploitation tended to perpetuate the pattern of semi-processing prior to export, they also provided potential raw materials for domestic industry.

Political and foreign influences

An important complex of political, economic and social factors also provided an impetus to industrialization. Intellectual arguments favouring industrialization were strengthened by external influences from the world economy. Dependence on primary exports and free trade became more vulnerable with fluctuations and declines in commodity prices, particularly affecting those countries heavily dependent on only one or two commodities. The impact of the First World War has generally been argued to have been a positive stimulus to industrialization, particularly in Argentina, Brazil, Chile and Mexico, cutting off markets and reducing manufactured imports and thus prompting domestic manufacture. In recent years this argument has been questioned, with evidence to suggest that though output, using spare capacity, increased to meet some scarcities, the amount of new investment and long-term expansion was limited.

The Great Depression played a major role in stimulating industrialization. It clearly revealed the vulnerability of export-dependency and the case for economic diversification and greater self-sufficiency, incorporating a more substantial manufacturing sector. This prompted moves towards import substitution industrialization (ISI), Latin America's first substantive move towards more autonomous industrial development. This process started in the 1930s, was heightened during the Second World War and continued in the 1950s, particularly after its precise articulation by Raúl Prebisch and the United Nations Economic Commission for Latin America.

These trends were closely linked to emerging economic nationalism, in which the republics sought greater control over the nature of their development, and greater independence from the perceived influences of Britain and the United States. Nationalist governments such as those of Getúlio Vargas in Brazil (1930–45), Lázaro Cárdenas in Mexico (1934–40) and Juan Perón in Argentina (1946–55) were influential in this process.

Such trends saw increasing direct or indirect involvement by the state in the economy. Governments began to provide tariff protection for consumer goods industries such as textiles, clothing, food and household goods, and concessions were granted on the import of machinery and essential raw materials. Government loans became available, and industrial banks to provide financial support were established. In Mexico the Nacional Financiera SA, created in 1934, provided funds for infrastructural and industrial projects, including the Monclava steelworks.

However, not all of the new initiatives came from government. Foreign companies continued to maintain interests in plantation agriculture, energy and transport, and mining, but also began to invest in manufacturing. This frequently formed part of an evolutionary process, from the export of finished goods to Latin America, through bulk import and local packaging, to local assembly of imported inputs or local raw materials and components. The automobile industry was a classic example of this process. Whereas cars had previously been imported into the major markets, from the 1930s the leading American firms such as Ford and General Motors began to export components for assembly in factories in Brazil and Argentina.

Although such foreign investment was limited in scale, in comparison with public and private domestic capital, it tended to concentrate in 'modern' industries, such as vehicles, chemicals, pharmaceutical and electrical goods, adding a new dimension to the continent's manufacturing sector. American and British firms were predominant, but French, German and Italian capital was also significant. The firms involved, which would now be termed multinationals, included ICI, British American Tobacco, Du Pont, Bayer, General Electric, Lever Bros, Pirelli, Armour and Nestlé.

Geography of manufacturing, circa 1945

By the late 1930s four broad elements were evident in the industrial structure of Latin America:

1. Processing industries linked to agricultural and mineral exports remained important, often associated with foreign investment and capital intensive technology.
2. Consumer-durable industries of engineering, vehicles and chemicals provided a new dimension, and the ISI process stimulated the expansion of the domestic industries of food, textiles and building materials.
3. Artisan activities survived in the small towns and rural areas, to meet local needs for food-processing, clothing, furniture and building materials.
4. In a few countries an additional element derived from the establishment of heavy industry, where natural resources, markets and levels of development were sufficient. By 1944 Brazil was producing almost 225,000 tonnes of steel and Argentina and Mexico over 100,000 tonnes; for most, however, industrialization was restricted to export processing and the production of consumer goods, such as beer (Table 3.1).

Table 3.1 Heavy and consumer good industries, circa 1945

County	Steel production 1944 (000 ingot/tonnes)	Beer production 1945 (000 hectolitres)
Mexico	175	3,687
Brazil	221	3,139
Argentina	150	2,575
Colombia		1,226
Cuba		780
Chile		684
Venezuela		401
Peru		345
Panama		262
Ecuador		257
Bolivia		214
Uruguay		210
Puerto Rico		159
Guatemala		79
Dominican Republic		35
Paraguay		31
El Salvador		24
Nicaragua		18

Source: based on data from Mitchell (1983)

By the late 1940s, there were considerable contrasts in the level of industrialization, its structure and spatial patterns. As Table 3.2 indicates, Brazil, Argentina and Mexico had the largest workforces, with Colombia, Chile and Cuba forming a second group. Most of the remaining republics had less than 100,000 workers. Only in Mexico, Venezuela, Peru, Bolivia, Chile, Argentina, Uruguay and Brazil did 'industry' (which included mining and public utilities) contribute more than 20 per cent of gross domestic product (GDP).

Table 3.2 Employment in industry in about 1950

County	Employed (000)
Argentina	1,457
Brazil	1,310
Mexico	998
Colombia	461
Chile	410
Cuba	336
Ecuador	234
Venezuela	178
Bolivia	151
Guatemala	113
Puerto Rico	110
Haiti	86
Jamaica	78
El Salvador	76
Paraguay	69
Dominican Republic	59
Nicaragua	39
Honduras	39
Costa Rica	32
Panama	20
Peru	no data
Uruguay	no data

Source: based on data from Mitchell (1983)

It is possible to identify how, by 1945, countries were in three broad groups, in terms of their industrial experience:

1. The more industrialized countries of Brazil, Mexico and Argentina.
2. An intermediate group which had experienced some industrial growth.
3. The least-industrialized areas of Central America, the Caribbean and Paraguay.

Within the individual countries there were also considerable contrasts in the scale and structure of manufacturing, with the national picture replicating the continental one. There tended to be a few centres of substantial and diverse activity, and an extensive periphery where industry was much less developed, smaller in scale and mainly limited to simple processing for local needs. Within this periphery there were a few enclaves of larger-scale, higher technology activity, often foreign-owned and concerned with export-commodity processing.

Argentina, Brazil and Mexico

By 1939 Argentina had over 54,000 industrial establishments, employing over 700,000 people. The food, drink and tobacco, and textile industries accounted for 36 per cent of jobs, followed by engineering and metal-working. Two-thirds of capital was Argentinian, but foreign investment was significant in certain sectors, such as the *frigoríficos* and in the new vehicle, electrical and chemical industries.

The spatial concentration of industry was marked. The 1935 census showed that over 70 per cent of the country's industrial wages were paid in Buenos Aires and its adjacent province. The capital, with one-third of the population, was the principal market and source of skilled labour, the leading port, the focus of the transport system and the seat of government. Not only was it the main industrial centre, but it was particularly attractive to the new growth industries and to larger factories. In 1935 47 of the country's largest plants were in the city and province of Buenos Aires, with seven in Santa Fé and two each in Cordoba and Entre Ríos.

The pattern of Brazil was remarkably similar. In 1940 there were 41,000 manufacturing plants and 815,000 employees, with food, drink and tobacco, and textiles and clothing providing almost 60 per cent of jobs. Figure 3.6 reveals that consumer industries dominated the pattern of employment. In all but one of those states which had more than 10,000 industrial workers in 1940, textiles and clothing, and food, drink and tobacco accounted for over half of the jobs. Within these industries there were some regional specializations, with cotton textiles and sugar processing being important in the north-east, tobacco growing in Bahia, and meat processing in Rio Grande do Sul. Raw materials similarly account for the other industries generating more than 10 per cent of state employment – iron ore in the metallurgy in Minas Gerais and pine forests in the wood-working of Parana and Santa Catarina.

Though there were more foci of industry than in Argentina, 66 per cent of industrial employment in 1940 was in the city and state of Rio de Janeiro, and in São Paulo and Minas Gerais. Away from the emerging industrial 'core' of the south-east, development was much more limited. The total industrial labour force in the interior, of Amazonia and the centre-west region was less than 20,000.

São Paulo, and not the then capital city, Rio de Janeiro, was the leading industrial centre. Prosperity from coffee had fostered its progress, and by 1940 São Paulo state had come to dominate the country's industrial geography, with over one-third of employment. Modern industries were strongly concentrated in the state: it had over two-thirds of the workers in the engineering and electric industries, and one-third of those in vehicles and chemicals and pharmaceuticals.

Foreign capital was significant in these 'newer' industries, and accounted for 27 per cent of industrial investment. The federal govern-ment had, however, become more involved in the economy than was the case in Argentina, in some areas of primary production and provision of public utilities. During the Second World War it also

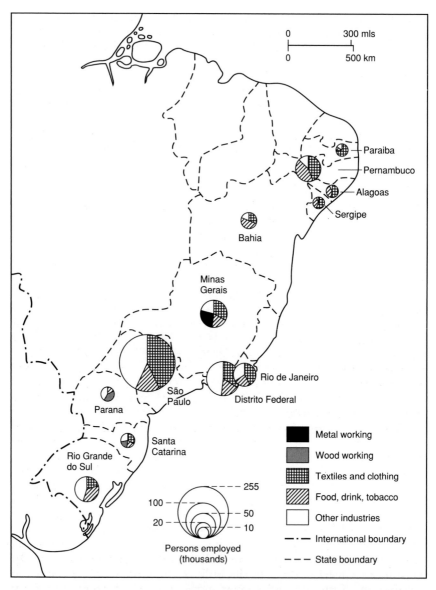

Fig. 3.6 Industrial employment in Brazil, 1940. Data relate only to workers, not managers, administrators, etc. Only states with more than 10,000 workers are shown. (*Source*: based on data from IBGE 1950)

became directly involved in the production of iron and steel, chemicals and vehicles.

In Mexico, following the revolution of 1910–17, the government had given high priority to agrarian reform, but by the late 1930s it began to provide financial and infrastructural support for industry. By 1940 there were almost 330,000 workers in mining and manufacturing. The former was of particular significance, both directly and as a basis for industrialization, with over 16 per cent of the workforce in the metal-mining and oil industries. In the manufacturing sector *per se* textiles and foodstuffs were the leading industries.

Minerals also served to give some dispersal to the location of industry, with oil refineries in the Gulf oilfields, at Tampico and Minatítlan, and lead, copper and zinc smelters in San Luis Potosí, Zacatecas and Coahuila. However, Mexico City and the adjacent Valley of Mexico were the dominant focus of the country's industry.

The intermediate group

Manufacturing activities in this group tended to be smaller in scale and later in date than those of the three leading countries, though there were some features in common. Chile provides a good example of spatial and structural contrasts in its industrial pattern. In 1940 its principal manufacturing industries employed some 116,000 people, with food and textiles as the leading sectors.

Following the decline of the nitrate industry the state established the Corporación de Fomento de la Producción in 1939, to encourage development. In addition to assisting the chemical, pharmaceutical, metal-working and tyre industries, the Corporación had provided half the capital for the Cia. de Acero del Pacífico steelworks at Huachipato, opened in 1950. It was intended to make Chile self-sufficient in iron and steel and to provide an industrial growth point in the centre-south of the country, dependent mainly on Chilean raw materials and sti-mulating local steel-using industries.

Minerals also provided a characteristically Latin American dimension to the country's industrial geography. Exploitation of low-grade copper ores required new technologies and capital from the developed world. This involved the massive mines of Chuquicamata, El Teniente and Potrerillos, developed by the American Kennecott and Anaconda companies. In addition to the mines, nucleations of processing plants, service activities and townships, sustained by external linkages and demand, were established in this otherwise inhospitable environment.

However, the principal concentration of industry occurred around the capital Santiago and its port Valparaíso, as the major attraction for ISI consumer industries. By 1945 the bulk of the food, drink and tobacco, clothing and furniture and similar industries were located around the Santiago–Valparaíso axis.

Similar patterns can be identified elsewhere. In Peru the capital/port of Lima-Callao attracted most of the consumer industries. Only the pull of raw materials offered a modification of this pattern. In the Andes the construction of the Lima–Oroya–Cerro de Pasco railway between 1870 and 1914 had permitted the mining of copper 4,000 metres above sea level. In the early 1940s the American Cerro de Pasco Corporation was operating five mining camps, three concentrating mills, a smelter, a coal mine and coke ovens, four power stations, brickworks, three chemical plants and 176 kilometres of railway, a classic example of a foreign enclave economy activity.

Colombia and Venezuela had made less industrial progress, constrained by small markets, low income, difficult terrain and limited transport. Oil production had provided Venezuela with a potential base for prosperity, and some refining capacity was established on the Lake Maracaibo and eastern oilfields, but much oil was exported as crude. During the Second World War the exploitation of iron ore in Venezuelan Guyana by the Bethlehem Steel Corporation laid the foundations for a later growth point. However, mineral-based activities aside, the focus of the country's limited manufacturing activity was Caracas.

Colombia, by 1942, had only 1,400 industrial establishments, two-thirds of them employing less than ten workers, which were mainly concerned with consumer goods. Colombia was atypical of the spatial pattern of Latin American industry, with a degree of dispersion, and lesser significance for the capital city. In 1945 Medellín was the leading industrial centre, with 22 per cent of jobs, followed by Bogotá (17), Barranquilla (10) and Cali (7 per cent).

Industrial activity in Ecuador and Bolivia was limited in scale and variety. In the former, over a third of an estimated 10,000 industrial workers were employed in the textile industry alone, with the food and drink industry as another major employer. The capital, Quito, and the main port, Guayaquil, were the principal industrial centres, with some dispersion deriving from oil refining on the Santa Elena oilfield, and from craft industries such as wool textiles and leather goods.

Limited development, lack of diversity and metropolitan dominance also characterized Bolivia. Tin mining provided over half of government revenue, and production, in the Potosí, Oruro and Uncía areas, was controlled by three large corporations. However, the ore was not smelted within Bolivia and there was little industrial spin-off. Other activity was limited to food, drink, textiles and clothing, with two-thirds of output coming from La Paz.

The least industrialized countries

The remaining countries have in common small size (all, except Paraguay, being below 130,000 square kilometres), small populations (all, except Cuba, having less than 3 million inhabitants *circa* 1950), and a marked dependence on a few export commodities, which generated little local processing activity.

Small size implied a restricted resource base, and low population and low income were major constraints on industrial development. In 1950 the six Central American Republics had a combined population of less than 8 million, and their economies were dependent mainly on coffee and bananas. The latter had been developed as a plantation crop by American corporations from the 1890s, with the companies transforming the fertile Caribbean lowlands, by creating plantations, drainage canals, railways, company towns and ports. Such developments created prosperity, but also acute dependency on a single crop and often a single foreign company, giving rise to the pejorative term 'banana republic'. In consequence, activity was limited to artisan industry and food processing, and confined to the main cities of San Salvador, San José, Managua, Tegucigalpa and Guatemala City.

Uruguay was also dependent on agricultural exports, with wool (40 per cent) and frozen, chilled and canned beef (20 per cent) dominating exports. The latter contributed a significant industrial element, with three foreign and one state-owned *frigoríficos*, and the Liebig meat-extract plant at Fray Bentos. Pastoralism also provided the raw materials for woollen mills and leather goods. Apart from the meat-processing plants, the markedly primate capital, Montevideo, dominated the location of industry.

Paraguay was the poorest, least populous and least-industrialized part of South America in 1945. Limited resources and debilitating wars in 1860–75 and 1932–35 provided little base for industry. Commercial agriculture was concerned with sugar, cotton and livestock, and the export of yerba maté and quebracho extract. Such industry as existed was linked to these products – cotton gins, two cotton mills, eleven sugar mills, a flour mill, four quebracho extract plants and three meat product factories!

Conclusion

By 1945 there was considerable range in the level and nature of industrialization in Latin America. The potential role of manufacturing in economic development had been recognized, but progress was variable. In the terminology of economic growth current in the post-1945 period, countries such as Argentina, Brazil and Mexico had achieved a 'take-off', in which rapid industrial growth was significant. Such success was in advance of that achieved elsewhere in the Third World. These countries had developed a fairly substantial and diverse industrial structure, with a range of consumer durable and non-durable goods, and some heavy industry. Others had begun the process of import-substitution, but in some parts of Latin America industry was limited to modest production of foodstuffs, clothing, and building materials.

Throughout the continent, though at varying scales, export-processing activities, based on agriculture and mining, perpetuated a pattern which had begun in the colonial period. These were geared to foreign markets, often foreign-owned, and using foreign technology. Craft and artisan industries, also with roots in the early colonial period, still provided everyday necessities in the towns, and especially in the rural areas, protected by distance from the competition of European manufactured goods. Only a few countries had begun to develop heavy and 'growth' industries, such as steel, vehicles, pharmaceuticals and electrical goods. The pattern of sources of capital which had evolved was a mixture of domestic private investment, foreign capital in the 'traditional' export and 'new' growth industries, and state capital in what were perceived to be 'key' industries.

Manufacturing activity was markedly concentrated in the capital cities and chief ports, or in centres such as São Paulo and Medellín. Away from these cores, export-processing industries created 'enclaves' in the rural areas or harsh environments, from which other manufacturing was entirely absent or limited to the simple processing of foodstuffs and other necessities. This pattern of concentration was to become of considerable significance in the post-1945 period, in terms of urban growth, patterns of migration, and regional development strategies. The 'modern' industrial sector remained dependent on imported technology. This, designed in the developed world, was increasingly capital-intensive, requiring less labour to produce a larger output of goods; yet for Latin America in this phase of industrialization, capital was scarce and labour cheap and abundant. In consequence, whatever contribution manufacturing might make to economic growth and diversification, its ability to absorb an increasing and underemployed population was restricted, a paradox which was to become of increasing significance in the modern period.

Further reading

Bakewell P (1984) 'Mining in colonial Spanish America', in **Bethell L** (ed.) *Cambridge History of Latin America*, Vol. II, pp. 105–51. Cambridge University Press, Cambridge. (A detailed analysis of history, technology and organization.)

Bell S. (1993) 'Early industrialization in the South Atlantic: political influences on the *charqueadas* of Rio Grande do Sul before 1860', *Journal of Historical Geography* **19**, 399–411. (A detailed account of the factors influencing the early development of meat-processing activities in southern Brazil.)

Brading D A (1984) 'Bourbon Spain and its American empire', in **Bethell L** (ed.) *Cambridge History of Latin America*, Vol. I, pp. 389–439. Cambridge University Press, Cambridge. (Includes discussion of trade, exports and the domestic economy in the eighteenth century.)

Crossley J C (1976) 'The location of beef processing', *Annals of the Association of American Geographers* **66**, 60–75. (A theoretical and empirical study of the evolution of the meat trade between Europe and the Pampas.)

Dean W (1969) *The Industrialization of São Paulo, 1880–1945*. University of Texas Press, Austin. (A comprehensive case study of the origins of Paulista industry and its capital sources.)

Dickenson J P and **Delson R M** (1991) *Enterprise under Colonialism: A Study of Pioneer Industrialization in Brazil 1700–1930*. Institute of Latin American Studies University of Liverpool Working Paper No. 12. (Explores the proto-industrialization of Brazil.)

Furtado C (1970) *Economic Development of Latin America: A Survey from Colonial Times to the Cuban Revolution*. Cambridge University Press, London. (A thorough overview of the development process in Latin America. Probably still the best introduction to the topic.)

Glade W (1986) 'Latin America and the international economy, 1870–1914', in **Bethell L** (ed.) *Cambridge History of Latin America*, Vol. IV, pp. 1–56. Cambridge University Press, Cambridge. (Provides a useful overview of sources of land, labour and capital.)

Haber S H (1992) 'Assessing the obstacles to industrialisation: the Mexican economy, 1830–1940, *Journal of Latin American Studies* **24**, 1–32. (A case study of the process and problems of Mexican development.)

Lewis C M 'Industry in Latin America before 1930', in **Bethell L** (ed.) *Cambridge History of Latin America* Vol. IV, pp. 267–323. Cambridge University Press, Cambridge. (A broad economic history from 1850.)

Libby D C (1991) 'Proto-industrialisation in a slave society: the case of Minas Gerais', *Journal of Latin American Studies* **23**, 1–35. (A detailed exploration of domestic industry in nineteenth-century Brazil.)

Miller R (1993) *Britain and Latin America in the Nineteenth and Twentieth Centuries.* Longman, London. (A first-rate overview of British investment and trade links with Latin America.)

Some of the above references are sources of data cited in the text.

CHAPTER 4

Environmental issues and the impact of development

Peter Furley

Introduction

The environment is perceived, measured and assessed in human terms. The physical, chemical and biological elements of environment are considered as *resources for* or *constraints on* human progress. In this chapter, the balance is explored between resource exploitation on the one hand and the buffering capacity of the environment to absorb disturbance on the other. South and Central America have the unusual distinction, within tropical and sub-tropical latitudes, of retaining a greater undisturbed proportion of their environment – though the rate of exploitation and impact is fast levelling the balance.

Latin America has been described as the 'land of promise and paradox' (WRI 1990). It possesses a rich array of natural resources (for instance oil and minerals, abundant sources of water, high primary productivity and biological diversity and some stretches of fertile soil), but these resources are unequally distributed and many regions suffer severe resource constraints. Similarly, it can be argued that the human resources show immense potential but, at the same time, much of the population lives in extreme poverty amidst increasing pollution and environmental degradation. Unchecked continuance of the trends identified in this chapter is almost certain to lead to decreases in biological productivity whilst threatening human health and endangering the very existence of myriads of plant, animal and micro-organic species.

Environmental diversity

The diverse environments of Latin America and the Caribbean present a palimpsest of current and past climatic patterns. These patterns have fashioned the present-day landscapes and ecology, which are further shaped by the distinctive structures and lithologies.

Climate and environment

The climatic regimes are dominated by oceanic influences but their variety reflects the latitudinal spread, nature and movement of atmospheric pressure cells, associated wind systems and the height and direction of mountain systems. The continental air masses typical of North America do not develop to the same extent over the smaller-scale,

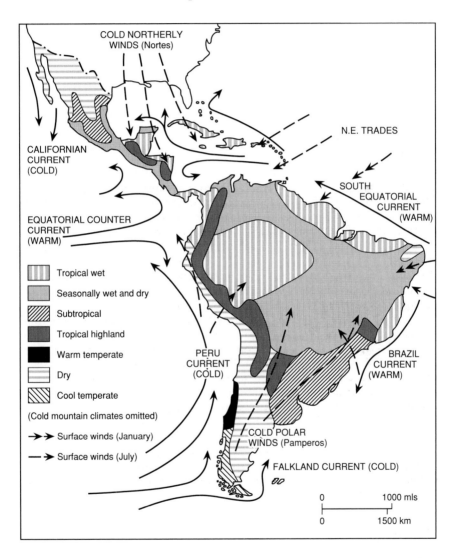

Fig. 4.1 Climatic types, ocean currents and wind systems: TW tropical wet (Köppenx); TW/D tropical wet and dry; ST sub-tropical; H tropical highland; WT warm temperate; D dry; CT cool temperate (*Source*: based on Collier *et al*. 1992; Nieuwolt 1977)

mid-latitude breadth of South America, whilst the only areas remote from the sea in Middle America lie in northern Mexico. Tropical and temperate systems clash and interact, although more than three-quarters of Latin America lies within the tropical belt. Depressions originating in the southern Pacific can produce cold waves and polar winds which may occasionally reach as far north as the Amazon but more frequently cross over the Argentinian and Paraguayan Chaco (pamperos) (Fig. 4.1). Such cold air masses inflicted great damage on the Brazilian coffee crop in 1981. Similarly the *nortes* of Mexico move southwards across Central America and may bring cold, crop-endangering winds. Seasonality is a powerful environmental determinant, although the relatively small area of land within the sub-tropics means that monsoons on the scale of India and South-East Asia are not a feature. However, periodic climatic displacements, such as the El Niño and Southern Oscillation events, originally thought to affect only the Pacific coasts, have been shown over the past 20 years to have both a regional and even global effect (see Box 4.1 and Fig. 4.2).

Ocean currents and associated wind systems also affect more local climates. The Pacific Humboldt and Californian cold currents travel from cold to warmer latitudes but their influence is enhanced by off-shore winds which pull cold, dense, nutrient-rich water from depths

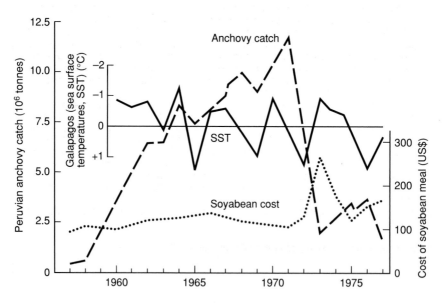

Fig. 4.2 Upwellings and fisheries, showing the economic impact of the 1972–73 El Niño event. The fall in the anchovy catch was shortly followed by a sharp rise in the cost of soyabean meal; this has been linked to large-scale clearance of Amazon forest for agriculture (*Source*: after Enfield 1992, based on Barber 1988)

Box 4.1:   El Niño and Southern Oscillation (ENSO)

ENSO events form the largest single global source of inter-annual climatic variation, with extensive and frequently damaging consequences over parts of South America and throughout the world. El Niño is the name given to abnormal oceanographic and atmospheric conditions which tend to occur around Christmas (hence the name) along the coasts of Peru and Ecuador in the upwelling zone. They are the result of slowly evolving, coupled interactions between the ocean and the atmosphere.

El Niño events were recorded as long ago as the end of the nineteenth century but became clearly identified in the 1920s. They have been associated with failures of the Indian Monsoon; dry periods and fire incidence in the Indonesian rain forest; floods on the Nile going back to at least the seventh century as well as droughts and storms in different parts of Middle and South America. Current research is linking evidence of climate anomalies with high resolution proxy events from ice cores, tree rings, coral records and varved sediments.

General characteristics: There are frequently warm (El Niño) and cold (La Niña) phases, but individual years may show markedly different patterns and there is no consistent definition of what happens. We do know that atmospheric pressures vary reciprocally between the south-east tropical Pacific and the Australian–Indonesian region, so that when one is lower the other is higher than normal. Usually the sea surface temperatures (SSTs) of the central and eastern Pacific off the Peruvian and Ecuadorian coasts are anomalously low for the latitude, a result of strong oceanic upwelling associated with the trade winds and the equator-ward flow of the Humboldt current. These low temperatures stabilize the lower atmospheres and inhibit rainfall, giving the hyper-arid conditions of the coast and a wedge stretching out into the Pacific.

Every few years ('aperiodicálly') this arrangement is disrupted. Heavy rain occurs and can last for months, associated with startling rises in SSTs from 2 to 8°C. What has happened is that the South Pacific High Pressure has weakened and thereby caused a decline in trade wind strength. This, in turn, weakens the upwelling of cold water allowing surface water temperatures to rise. Such oceanic warming leads to evaporation and heating of the troposphere leading to the instability and convective storms already mentioned. Climatic records show that this has happened on many occasions in the past.

Effects: Within Latin America, the most significant effects are the heavy rains and flooding over Peru and Ecuador, sometimes extending south to include Central Chile. Such climatic changes have been associated with droughts in north-east Brazil and parts of Central America and also to coral mortality in the eastern Pacific reefs, where coral bleaching (whitening due to the death of the endosymbiotic algae) has led to 95 to 98 per cent mortality in some localised areas. The ENSO events also contribute to the lowered marine productivity off Peru (see Box 2).

In recent years, the worst events have been in 1972–73, which had severe economic effects in Peru, and 1982–83 when there were torrential floods and landslips in northern Peru and Ecuador with extreme drought in southern Peru and Bolivia.

Sources: Diaz and Markgraf (1992); Glynn (1990); Philander (1990).

and give rise to the distinctive biologically rich upwellings (see Box 4.2). Over the Atlantic coasts, the opposite effects occur with warm, wet winds blowing off the ocean and across the Amazon Basin and Central American coast.

Box 4.2:   Upwellings and the Peruvian fishing industry

The vícissitudes of the Peruvian fishing industry highlight the interactive roles of natural processes and human activities. Through a combination of El Niño events (Box 4.1) and overfishing, the industry had plummeted by the end of the 1980s, to less than 50 per cent of its peak in the early 1970s and the species composition had noticeably changed (Fig. 4.2).

The upwelling systems pull cold, nutrient-rich water from the ocean depths which stimulates high primary productivity and so encourages a food chain giving the rich anchovy (*Engraulis ringens*) fishing grounds of Peru. These in turn support the sea birds which have left their economic mark in the guano deposits used for many years in phosphate production. This was an example of dynamic equilibrium but one jolted periodically by the ENSO events outlined earlier. In normal circumstances the equilibrium would be restored, but in the case of Peru the pronounced events of 1972–73 and 1982–83, together with overfishing, have led to a precipitous drop in the catch with very slow recovery.

Upwelling only occurs over about 0.1 per cent of the oceans but is estimated to contribute 50 per cent or more to the world's fish harvest. The Humboldt current and its upwelling system is one of the richest of all marine ecosystems, with high densities of phytoplankton and zooplankton. In the 1960s and 1970s it represented around 20 per cent of the world's fish catch, according to Caviedes and Fik (1992). In 1972 there was a particularly rich anchovy harvest, later found to have been trapped inshore by the warmer waters of an El Niño event. Once the dense schools of fish had disappeared, the catch (previously at 170,000 tonnes a day) dropped to around 2.5 million tonnes over the whole season.

Before the 1950s then, the Peruvian fishing industry was small and served national needs. During the 1950s fish meal was produced for export along with fish oil. By the 1960s the industry grew so fast that it was the world's leading producer with a fleet of around 1700 boats. Although it was recognised that the trend was over-exploitative even during the 1960s (a maximum sustainable yield was estimated to be 9.5 million tonnes per year, which includes 2 million tonnes consumed by sea birds), the industry was unrestrained. It was this combination of political misguidedness, concentration on a single species, commercial greed and the ENSO events which led to the near collapse of the industry.

Sources: Pauly *et al.* (1989); Barber and Chavez (1983); Caviedes and Fik (1992); Glantz and Thompson (1981).

Climate and vegetation formations

A number of distinctive bioclimatic regimes may be identified which help to characterize the rich, ecological and environmental individuality of the region.

Middle America

The outstanding environmental feature of the connecting isthmus from Mexico to Panama and the linked arcs of the Caribbean is its great diversity. With spectacular island chains, active volcanic ranges, plateaux and striking mountain systems, flat limestone plains, distinctive river basins, together with a multitude of lakes, the manifold landscapes have been an appropriate stage for the historical pageant of peoples who have settled the region.

As a result of this physiographical and geological complexity, the ecology is a complete contradiction of the widely held notion that tropical environments are extensive, rather homogeneous tracts. There are both temperate and tropical elements to the vegetation. The deciduous and evergreen temperate forests are found over the highlands, running down the axes of the structural cordilleras. At lower altitudes, there are semi-humid and semi-arid grasslands similar to those found in South America. Scrub forests with evergreen shrubs form low forests in Mediterranean climates and also small pockets neighbouring the arid scrub of desert margins. More typically, tropical vegetation predominates south of the Mexican plateau and on the Caribbean islands. Moist forest is still found, though heavily disturbed in lowlands under 300 metres. Where there is a more pronounced dry season, deciduous forest is widespread with savannas and dry scrub forming tropical counterparts to the warm temperate grasslands.

Amazon Basin

The Basin contains the world's greatest volume of fresh water flowing to the oceans (about 20 per cent of global outflow), and is characterized by westward-flowing, warm and damp air from the Atlantic, driven by trade winds and with frequent thunderstorms. It is often thought that the Amazon lacks seasonality, and this may be true towards the Andes where increasing height provokes orographic rainfall. Most of the Basin, however, is typified by marked dry periods with intense evapotranspiration (Fig. 4.3). Spells of even a few days and certainly a few weeks, can cause powerful drying and provide an opportunist 'burning season'. The impact of forest clearance has been shown to have a potentially devastating effect (see Box 4.3). The true evergreen forest reflects the areas of more continuous rainfall, but the Basin is surrounded by concentric rings of more semi-evergreen and semi-deciduous forest. Savannas take over where the seasonal dryness is too pronounced for tree growth to be maintained. Natural and anthropogenic factors shape this ecological 'tension zone', where the dynamic forces controlling the two major ecosystems meet at a constantly changing boundary. It is not often realized that some 20 per cent of *Amazônia Legal* within Brazil is non-forest – made up mainly of savannas or swamps in a mosaic of varying plant formations.

Both the Atlantic forests on the Brazilian and Central American coasts and the Colombian and Ecuadorian coastal forests flourish in similar conditions of onshore warm, wet winds and provide some of the richest and most beautiful forests in the Americas.

Montane ecosystems

Mountains generate more localized climates. They act as barriers to wind systems causing dry zones on the leeward sides, as on the Pacific side of the Andes.

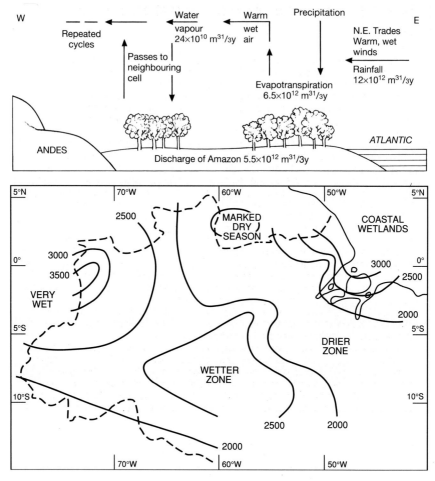

Fig. 4.3   (a) Rainforests and regional climate; (b) rainfall over the Amazon
Basin (*Source*: from Salati 1985)

Rising above the lowlands from around 1,000 metres, local climates
become cooler (*tierras templadas*) and forest may be replaced by savanna
where soils are poor and rainfall seasonal (typical of Belize, Honduras
and Nicaragua). Cloud forests develop where moisture-laden winds are
forced to rise (typical of Guatemala, southern Mexico, parts of
Venezuela, eastern Peru and Colombia). At these altitudes, seasonal
temperature variations are slight, averaging 15 ℃, although diurnal
variations are more extreme. At greater heights, cold temperate
conditions prevail and over 2,000–3,000 metres the tree line is reached
and montane shrub and grasslands predominate (such as the *paramos*),

Box 4.3: Impact of Amazon forest clearance on climate

There is continuing debate amongst scientists as to how far deforestation affects local regional or global climate. We do not know for certain the extent of disturbances to rainfall or potential desiccation but there are a number of signs which suggest that the potential damage is severe. Water balance is intimately related to energy balance, so that any model of the water cycle has to account for its influence on energy flow and vice versa. It is highly likely therefore, that the clearance of large tracts of forest will markedly interrupt the equilibrium.

The fate of westward-flowing warm, wet winds brought in from the Atlantic has been modelled by Salati and his fellow researchers. The incoming winds bring rain which sustain the forests to the east of the basin (Fig. 4.3a). The rainfall is largely convective and passes across the region in a series of 'cells', towards the west. These cells are regenerated by evapotranspiration from the forest canopy and ground surface, and the atmospheric moisture is then available to be passed on to the next cell. Using isotopic techniques for water vapour monitoring, it has been proposed that as much as 50 per cent of the moisture in the air flow may be derived from this process. It will be evident that if the forest is cleared, there will be less moisture available to neighbouring or downwind streches of forest (Fig. 4.3b).

The problem is augmented by the normally drier belt lying across the Middle Amazon (Fig. 4.3b). There are fears that the westward passage of moist air could be disrupted quite rapidly. However, it should be noted that there is no universally accepted view of the water quantities and flows since little data is available over this very extensive and remote area. Furthermore until recently, the level of error in assessing deforestation or land use change was so great that its relationship to climatic models could be predicted with much confidence.

Nevertheless, if these processes are correct, then the disruption could be felt outside the Amazon. At present, the moist air flows pass inwards from the Atlantic to the Andes, giving increasing rain as they rise towards the west. Not only is this rich forest under potential threat but, because the wind patterns are partially deflected obliquely back across Central Brazil, the process could eventually lead to water deficiences in what is one of the fastest growing food-producing areas in the Americas.

Sources: Molion (1987); Salati (1985); Henderson-Sellers (1985).

with alpine communities taking over up to the frozen zone over 4,000 metres (*tierra helada*) (Fig. 4.4).

On the other hand, the height of mountain systems throughout South and Central America means that snow accumulated in the mountains provides a constant source of water for the outflowing stream systems.

Ecosystems dominated by seasonality

Away from the semi-permanently wet mountains and forest areas, the increasing seasonality (especially the prolonged dry season which may extend to eight or even ten months) gives rise to deciduous forest or arboreal, shrub and grass savannas (Fig. 4.5). Fires are typical of such areas and the vegetation is characteristically fire-adapted. It would be inaccurate to talk of savanna climates, since these landscapes have evolved over several distinctive climatic regimes helped by soil, slope,

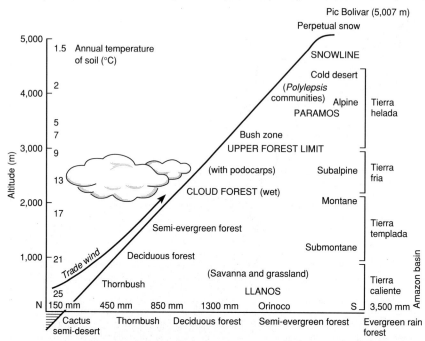

Fig. 4.4 The zonation of vegetation with altitude: north to south across Venezuela. In Venezuela, the vegetation communities along a north–south transect at sea-level range from semi-desert through a thorn bushland to deciduous, semi-evergreen and eventually evergreen rainforest in the Amazon Basin. With increasing altitude, the transitional zones of Latin America may be recognized – the Tierra caliente below about 1,000 m with a thorny savanna to Llanos grassland (from north to south), the Tierra templada from around 1,000 m to around 2,000 m with deciduous to evergreen to cloud forest on the seaward-facing slopes, the Tierra fria with an upper limit around 3,000 m which also represents the upper tree limit, and the Tierra helada reaching up through alpine grasslands (paramos) to the snow line at around 4,500 m. (*Source*: after Furley and Newey 1983, based on Walter 1973)

fire and evolutionary factors as well as the water regime (Box 4.4). Consequently, these are highly dynamic systems, constantly shifting according to human and environmental pressures.

Arid and semi-arid ecosystems

Several areas of South and Central America suffer regular drought or persistent dryness, whilst northern Mexico, coastal Peru and northern Chile have evolved extreme deserts. North-east Brazil and an oblique NE–SW line through Brazil to the Chaco provides a line of dry landscapes usually accentuated by calcareous outcrops, where a xerophytic vegetation has remained since the greater aridity of the Pleistocene.

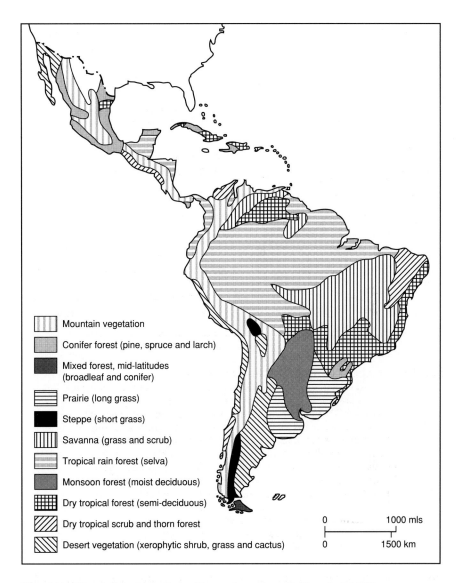

Mountain vegetation

Conifer forest (pine, spruce and larch)

Mixed forest, mid-latitudes
(broadleaf and conifer)

Prairie (long grass)

Steppe (short grass)

Savanna (grass and scrub)

Tropical rain forest (selva)

Monsoon forest (moist deciduous)

Dry tropical forest (semi-deciduous)

Dry tropical scrub and thorn forest

Desert vegetation (xerophytic shrub, grass and cactus)

0   1000 mls

0   1500 km

Fig. 4.5 Vegetation formations (*Source*: based on Collier *et al.* 1992)

These more semi-arid areas are typified by rainfall irregularity. The rainfall uncertainty, with more favourable periods alternating with drought, has threatened the stability of agriculture and settlement since the earliest days of colonization. Areas in the rain shadow of mountains are similarly affected, with typical föhn winds such as those striking the east of the Andes – for example in centre-west Argentina. *Caatinga* vegetation is typical of the plant formations in very dry areas. It is a microphyllous or small-leaved, thorny woodland 6–8 metres tall, which

Box 4.4: Pleistocene refugia and their environmental importance

At periods of maximum ice advance during the Pleistocene, climates are believed to have been only slightly colder but definitely drier throughout the humid tropical lowlands. Repeated oscillations of climate over about a million years, left their mark on the landscape and vegetation in a way which is still relevant today.

Although the areas to which forests may have retreated during cold phases is not precisely known or agreed, the broad idea of forest contraction and savanna advance has general support (Fig. 4.6). It appears that forests are naturally

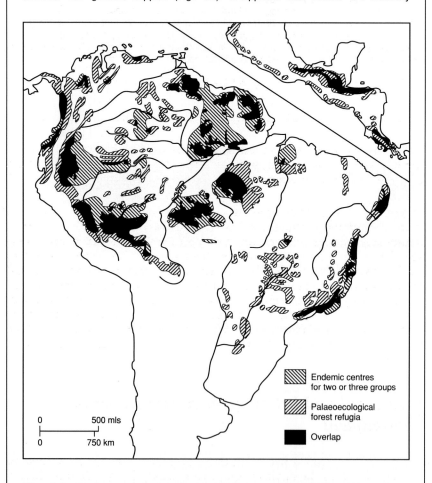

Fig. 4.6 Endemic centres and palaeocological refugia. The refuge model is based on palaeoclimate, topographic, geomorphological, soil and vegetational data. Note the appreciably larger endemic areas around the proposed refugia (*Source*: Whitmore and Prance 1987)

Box 4.4 Pleistocene refugia and their environmental importance
(continued)

expanding at their margins at the present time – a contrary force to that of deforestation, but important because it suggests that regeneration could be effective if given a chance.

It is valuable to understand these processes for several other practical reasons. Firstly, knowledge of the refugia clarifies understanding of the current distribution of wildlife which has radiated from the refugia during more favourable periods. Secondly, and working the other way round, scientists are trying to follow speciation (as of butterflies or amphibians) back to source areas. This enables conservationists to give better advice to governments concerning conservation and management policies. The logic behind this is that refugia should theoretically act as reservoirs for species diversity. There is little doubt that even within an apparently similar tract of vegetation, there are some patches which are species-poor and others which are spectacularly species-rich. The latter are valuable for their genetic, medical and other economic potentials.

Sources: Myers (1984); Prance and Lovejoy (1985); Whitmore and Prance (1987); Furley *et al.* (1992).

can withstand six or seven months with negligible rainfall. At the end of the dry season, the trees appears completely dead with a ground flora of lower shrubs, cacti, bromeliads and grasses.

Temperate grasslands and forests
The southernmost states in Latin America are affected by frontal depressions giving winter rain, although summer thunderstorms may also be an important influence on land use. In the extreme south of Chile, the climate is typically cool temperate with cloud and rain and moderate to cold temperatures, with increasing winds towards the inhospitable Tierra del Fuego. In northern Chile, a narrow band of warm temperate conditions is encountered with a Mediterranean climate similar to California. These systems seem to be more stable than their tropical counterparts and more resilient to environmental or human pressure.

Structural and lithological controls over the environment

Much of the historical evolution and development has been influenced and fashioned by the environment. The geological procession in Central America, for example, has been responsible for the intermittent connection between North and South America, helping to regulate the extinctions and dispersals of animals and people. Equally, it has strongly influenced the response of successive waves of immigrants. Geological composition as well as structure have been important, as illustrated by the determining role of karstic limestones in the Yucatán, which have affected the siting of settlements and the nature of Mayan agriculture. The spread of the conquistadors from Mexico into the rest of Latin America was similarly shaped by the north–south trending cordilleras (Fig. 4.7). The very intermixture of sea and land, so much a feature of Middle America, has been reflected in the movement and history of its peoples. This section, therefore, briefly portrays the

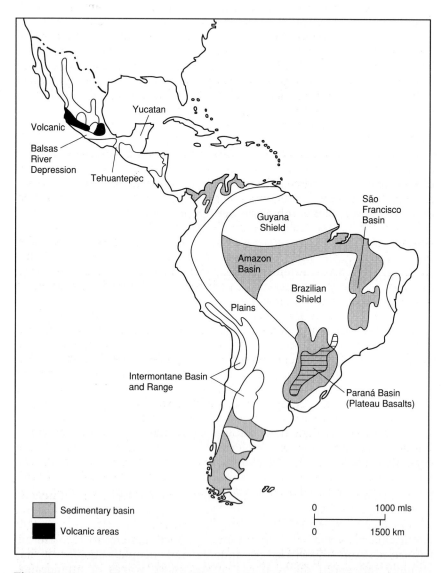

Fig. 4.7   Structural environments (*Source*: after West and Angelli 1966; Collier *et al*. 1992)

environmental variety which underpins much of the historical evolution.

The Mexican plateau is the most significant physiographical feature in Middle America, a tilted upland block reaching nearly 2,500 metres in places along the southern rim and dropping to less than half that height towards the north. The plateau surface, though appearing level, is made up of low residual mountains separated by basins, many of which are filled by shallow and fast-sedimenting lakes around which settlements have evolved. The southern part of the plateau is volcanic mostly of Late Tertiary age, with spectacular cones such as Popocatepetl (over 5,500 metres) or Orizaba to the east (5,754 metres). Many areas are still active or have erupted in recent times, such as the well-documented Paricutín (1943) and El Chichón (1981). The edges of the plateau are dramatic, the east comprising many elongated and highly folded limestone ridges (Sierra Madre Oriental), whilst the west (Sierra Madre Occidental) is volcanic and cut by deep canyons (*barrancos*).

South of the plateau separating northern Mexico from the rest of Central America lie a complex group of physical environments which have shaped the emergence of distinctive cultures. The Balsas River depression divides the plateau heartland of Mexico from a series of discontinuous, crystalline ridges forming the mountainous Sierra Madre del Sur. Towards the Pacific lies a narrow coastal plain often interrupted by cliffs which have inhibited settlement, stretching from the dry peninsula of Lower California in the north to Panama. The narrow isthmus of Tehuantepec has formed an important east–west corridor only 215 kilometres across. In contrast to the Pacific side, the Atlantic coast has wide plains extending most of the way down Central America but interrupted by swamps and large lagoons with occasional outliers of mountain ridges and volcanic hills.

The Caribbean islands rise from arcs related to the mainland structures. Geologically similar, their isolation and fragmentation and the fact that they faced the thrust of European colonization has fashioned a very different history. The complex area, known as Old Antillea, extends from Chiapas in south-east Mexico to the greater Antilles islands, consisting of east–west arcuate mountain ranges and depressions made of limestone and sandstone and forming very rugged, highly eroded landscapes. To the north of this area lies the foreland on which the island ridges have folded. It consists of a recently emerged limestone surface, one of the largest in Latin America, overlying a crystalline platform. The Yucatán forms a distinctive karstic landscape with sink-holes (*cenotes*) and numerous solutional depressions (*bajos*), both of which were vital for the Maya settlements which covered most of the area in the past.

The remainder of Central America is made up of faulted and volcanic mountains and, like the Old Antillean region, extends out into the Caribbean in arcs forming the Lesser Antilles. Numerous well-known volcanoes such as Fuego (3,877 metres) in Guatemala, or Irazú (3,477 metres) in Costa Rica, rear up dramatically when viewed from the Pacific and have closely interacted with the historical occupation of the region. This immense geological diversity is one of the reasons why Middle America contains such a wealth of history, cultures and varied landscapes.

By contrast, in South America, the environment is shaped by large-scale crustal plates resulting in broader morphological regions. Three main structural units can be identified. The first comprises the uplifted and warped faulted blocks of Pre-Cambrian age. This ancient platform is likely to underlie most of the continent. The land surface is made up of deeply weathered and dissected metamorphic and igneous rocks, the latter forming residual hills (inselbergs) in places. It is thickly covered by later Palaeozoic and Mesozoic sediments and has been subject to considerable faulting; for instance the southern boundary forms a fault scarp creating the Iguaçu falls. The second unit consists of downwarped basins within the Precambrian basement. Such basins have been infilled by later sediments and are, in places, over 4,000 metres deep. Most of the infilling dates from the rise of the Andes in the Late Tertiary and the establishment of the present course of the Amazon 8–15 million years ago. Sediments indicate a rate of deposition of 50–115 cm every 1,000 years over the past 2–3 million years. The third unit is made up of folded mountains dominated by the Andes and Caribbean coastal ranges. Within the Andean cordilleras, there are downwarped basins, notably those of Lake Titicaca and the Bolivian Alto Plano. These basins are filled with fluvial, volcanic and glacial deposits and furnish the only agricultural possibilities within the generally inhospitable landscape.

Human environments

Whilst natural forces predominate over most of the landscape, the influence of human activity, especially over the past 50 years, has created distinctive habitats with their own sets of environmental problems.

The anthropogenic environment is most clearly characterized by urban settlements. The concentration of population, buildings and industry has created an artificial environment with its own positive feedback loops which reinforce its impact on physical and biological resources. The mega-cities epitomize this trend and are considered later. It is sufficient here to note the impact on the atmosphere (as witnessed by air pollution, heat islands, wind channelling), on water quality (toxic sewage and waste, domestic and industrial pollution, temperature increases, disease and diversion of natural channels often leading to flash flooding, inadequate storm provision and altered velocities), and on biological resources (a highly reduced, polluted and often exotic flora and selected fauna which can associate with or benefit from dense human communities). Extremes are found in poorer areas – the *favelas* and *barrios* which typify virtually all Latin American cities.

The impact of human activities on the environment extends well beyond urban areas. The spread of pollution affects all the area and passes beyond national boundaries. Dust and gases travel the world and the impacts of industry and land degradation on the atmosphere are also global. Road construction has had a marked, highly localized impact, but one which traverses vast areas of the continent and spreads the virus of pollution through settled and sparsely populated areas. Dams and artificial water bodies create their own local atmospheres. Most pervasive of all, the clearance or disturbance of natural vegetation upsets the natural gas balance, adds gases such as carbon dioxide to the atmosphere through burning, increasing the decomposition of organic matter and release of nitrogen through ranching activities and, more obviously, by completely changing the nature of the biological patrimony taken for granted for generations. A number of these points are taken up later.

Environmental problems affecting human populations

The character of the environment poses numerous difficulties for human societies, whilst they in turn frequently exacerbate the problems through deliberate or unconscious mismanagement.

The aspects of the environment which constrain human activity can be divided for convenience into air, land and water.

Meteorological and atmospheric extremes

Some of the most dramatic of atmospheric problems consist of unpredictable storms and high winds, referred to as hurricanes in the Caribbean, where as many as 5–12 major storms may hit the region between July and October. They are characteristic in various forms in many parts of Central and South America. Often related are problems of intense rainfall causing physical damage to buildings and crops as well as through flooding. On many parts of the low-lying coast, wind force also brings in destructive storm waves; a wall of water 2–3 metres deep was believed to have passed over Belize City in the hurricane of 1932, with the loss of life of several hundred people. At the same time, large wooden vats storing domestic water at that time, were uplifted by the torrent and charged through the city destroying many of the fragile buildings. A further catastrophic hurricane hit the City in 1961 (see Fig. 4.8).

At the other extreme, too little rainfall is a feature of several parts of the continent as shown in the next section, and droughts are a recurrent plague of several areas such as the *sertão* or outback of north-east Brazil (Box 4.5). In these areas, high insolation during the day causes the highest temperatures in the region, whereas radiation back into space at night can lead to very cold conditions.

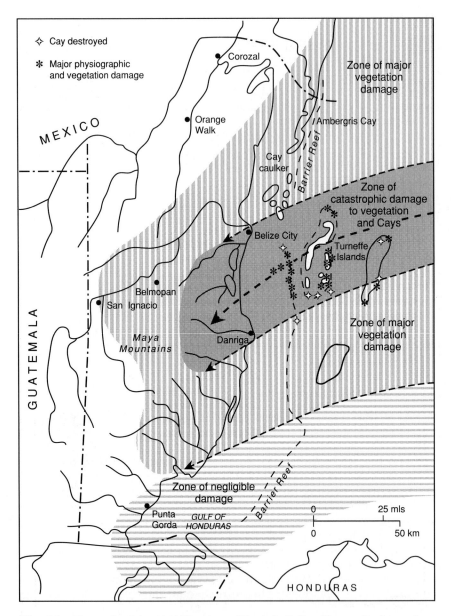

Fig. 4.8 The path of the 1961 Hurricane Hattie in Belize (*Source*: Furley and Crosbie 1974)

Box 4.5: Growing aridity

One of the environmental concerns of global importance is that of expanding aridity. Desertification literally implies enlarging the margins of deserts, but perhaps as serious and less obvious is the insidious desiccation of semi-arid areas and their progressive loss of biological productivity. It represents an extreme form of land degradation (Figs. 4.9 and 4.10).

Dry land vegetation and desert scrub are widespread in Latin America, representing around 10 per cent of the world area. Mexico contains around 100 million hectares, Argentina over 93 million hectares, Chile nearly 27.5 million

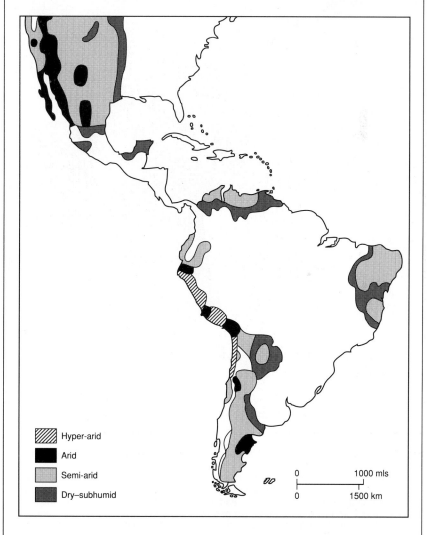

Fig. 4.9   Aridity (*Source*: after UNEP 1992)

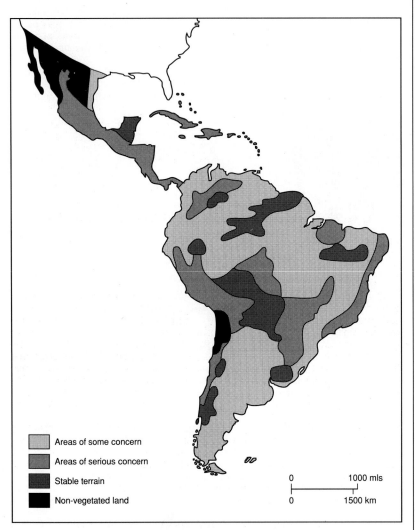

Fig. 4.10 Land degradation (*Source*: after WRI 1992)

hectares and Peru some 15.25 million hectares, with smaller but significant stretches in Venezuela and Bolivia. The only truly hyper-arid regions are those of the Atacama Desert in Peru–Chile and tracts of the Sonoran Desert in Mexico. Expansion from these core areas has been the focus of many studies, with concern for semi-arid vegetation formations such as the Brazilian *caatinga* or the Argentinian and Paraguayan *chaco*.

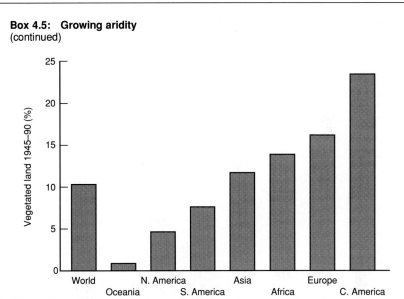

Box 4.5: Growing aridity
(continued)

Fig. 4.10 (continued) Moderate, severe and extreme degradation as a per cent of vegetated land 1945–90

Chile provides a good example of the desertification problem with a sharp gradient from the hyper-arid desert in the north. Water deficits are becoming serious and run-off losses are increasing as the protective vegetation cover and soil organic matter are removed. Salinization is also increasing in areas where groundwaters contain dissolved salts which are then evaporated at the surface. Rainfall records have not had a sufficiently long run to permit the conclusion that there is a gradual trend to a more arid climate; the observed rainfall decreases may be part of a longer global pattern. Nevertheless, human action is disturbing the natural vegetation through cultivation and overgrazing and has led to the invasion of less productive species, the rapid decomposition and mineralization of organic matter, greater run-off and soil erosion and weakened soil structures.

The problem of progressively worsening aridity is a consequence of natural vulnerability of marginal biological ecosystems together with inappropriate land use systems. It is possible to predict thresholds beyond which the resource degrades rapidly once its resistance or buffering capacity has been overcome. Chapter 12 of UNCED's Agenda 21 was devoted to desertification and drought, with an attempt to gain greater international recognition for concerted action in six interrelated programmes to combat desertification.

Sources: Thomas (1993); Thomas and Middleton (1994); Grainger (1990); Government of Chile (1980); UNEP (1993).

Earthquakes and volcanic activity

The instability or fragility of the land also poses problems. Latin America contains about one-fifth of the world's seismic activity and many mountainous areas are affected by earthquakes or volcanic eruptions (Fig. 4.11). Examples are found throughout the history of the

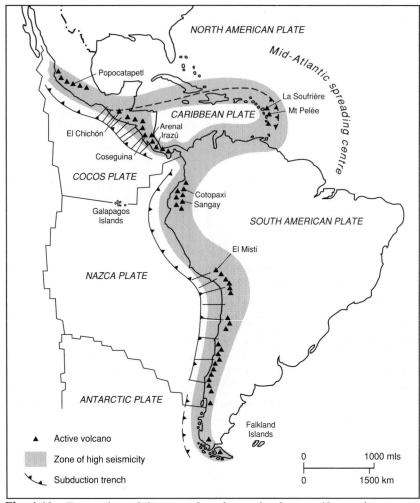

Fig. 4.11   Zones of instability – earthquakes and volcanoes (*Source*: from Collier *et al.* 1992: 25)

region, as in numerous events which led to the abandonment of Antigua, the old capital of what is now Guatemala, or El Salvador which was severely damaged on 14 occasions in its colonial history. More recently, disasters have occurred in Nicaragua (1972), when the capital Managua was badly damaged, and in Mexico City (1985). Earth tremors frequently initiate landslides and, if compounded with snow avalanches and ice melt, can precipitate disastrous floods and catastrophic mass movements. This is a frequent occurrence over the precipitous slopes of the Andean states; the Cordillera Branca in Peru is notorious for mass movements following tremors, and ancient flows still scar the hillsides. In 1941 such an event killed 4,000 people at Huares, with 3,500 at Ranrchirea in 1962 and a further 500 in 1970. Mud flows also result from super-saturation of unconsolidated sediments over impermeable parent materials, as in catastrophic slips in many of the Rio squatter settlements.

Volcanic eruptions have long been a part of life in some districts. The 1835 eruption in Nicaragua's Gulf of Fonseca damaged an area with a radius of 160 kilometres, whilst in 1902 the most violent form of eruption was named after Mt Pelée which killed 37,500 people in Martinique and a further 1,500 in adjacent La Soufrère. In the same year, the eruption of Santa Maria volcano in Guatemala accounted for 600 lives. Where igneous activity occurs in mountainous areas, there is the added danger of triggering snow melt, leading to avalanches and terrifying mud flows. All of the Andean countries have experienced tragic incidents of this sort, such as Villarica in Chile (1963), and some areas are known to be potentially very dangerous, such as El Misti in southern Peru.

The impacts vary greatly. Crops and fields may be physically overrun, as in the well-documented Paricutín volcano in Mexico (1943–52), or destroyed by deposits of dust and ash which can also affect whole industries. Examples include the destruction of the coffee industry in Costa Rica by eruptions from Mount Irazú and other lava and ash outpourings, between 1963 and 1965, or the spectacular eruption of El Chichón in Mexico, which not only caused great destruction and killed around 2,000 people, but also poured dust into the upper atmosphere – which was carried round the world for several years. At the same time, the paradox of fertile soils, often associated with the rapid weathering of magmatic lava deposition, means that human settlements have often been most dense in areas of potential instability and thereby occasion greater disasters when eruptions occur.

Soil constraints

Since three-quarters of Latin America is tropical, the constraints of tropical soils assume a major importance. Two broad groups may be considered, divided for simplicity into humid and arid areas.

Within the humid tropics, there are difficulties for human use posed by both well-drained and poorly drained soils. The poorly drained or hydromorphic soils (Table 4.1) are found in more obvious localities such as swamps, marshes, seasonally flooded riverine plains and mangroves, but are also widespread in shallow depressions over level surfaces. Despite the problems associated with utilizing saturated soils, even if only during wet seasons, including the lack of oxygen leading to reduction processes and anaerobism, the hydromorphic soils have been increasingly utilized for cultivation. They often have the advantage of being base-rich, since they are young soils usually from alluvial or lacustrine deposits. They vary markedly depending upon mineral provenance; for example, soils derived from sediments brought down from the Andes are usually mineral-rich and fertile, whilst streams draining off the old, weathered shield areas have little sediment, are nutrient-poor, often acidic and organic. Hydromorphic soils are used both for dry crops at appropriate seasons of the year and for wet crops such as rice, which is becoming progressively more important as an alternative staple to maize and beans. The problem lies in the difficulties of applying correct management to these soils and also to the fact that they are frequently sites of great biological productivity; in other words, of regional and often international environmental importance.

Table 4.1

Oxisols:	intensely weathered, old soils and their eroded accumulations; acidic with low fertility; often iron-rich and with toxic levels of aluminium.
Ultisols:	highly weathered soils with a horizon of clay accumulation and low base status.
Mollisols:	organic surface horizon, dark coloured with high base status.
Vertisol:	high content of swelling clays which shrink and crack in the dry season; dark coloured; often fertile but difficult to manage.
Entisols:	weakly developed, recent profiles without horizon formation.
Inceptisols:	young to moderately developed soils with rapid formation.
Aridisol:	dry and very arid soils; low organic matter.
Alfisol:	moist, sub-surface clay accumulation and moderate to high base content.
Andisols:	volcanic soils.

Amongst the soils which are seasonally very wet can be included the vertisols, which typically occur in flat, low-lying situations. They are dark clay soils which swell in the wet and shrink and crack in the dry season. They are often base-rich, derived from limestones and calcareous materials, as in the Yucatán and some Caribbean islands. They have high clay percentages, which are responsible for the swelling and shrinking. They are potentially fertile but extremely difficult to manage.

Most well-drained soils in the lowland forests and savannas require skilled management if they are to be used for agriculture. They are predominantly oxisols and ultisols and are deep, extremely weathered, acidic and low in nutrients. Acidity is often reinforced by (frequently toxic) levels of aluminium. Whereas native plants have devised mechanisms for overcoming these problems, introduced or exotic plants become highly stressed and need both skill and capital inputs to survive. Consequently, these areas are better suited to commercial farmers than smallholders. The latter are rarely able to get above a subsistence level as a result of the edaphic difficulties facing them. A further common problem concerns the process of plinthite formation, which is the accumulation of iron. Iron compounds may oxidize and harden to ironstone following vegetation clearance. A further group consists of volcanic soils, which vary from deep and base-rich when they are derived from intrusive ferro-magnesium parent materials, to acidic, poorly structured volcanic ash. As indicated earlier, some of the greatest densities of rural population in Latin America are to be found in close association with volcanic soils.

Arid zone soils are clearly limited by water deficits. They are potentially fertile, since they lack the leaching and chemical weathering of more humid areas. Semi-arid soils are particularly prone to further desiccation (see Box 4.5) and degrade rapidly if mismanaged. Such soils are characteristic of the desert areas of Latin America and of the semi-arid areas, notably northern Yucatán, north-east Brazil and the Chaco region forming the zone running south-east into Argentina.

Outside the tropical and sub-tropical environments, there are temperate soils whose character reflects latitude or altitude. In cool temperate areas, the more typical are alfisols and mollisols, ranging from excessively well-drained soils (alfisols) to seasonally moist (mollisols). Many of the management problems associated with these soils have been addressed by scientists working in analogous areas of North America and Eurasia, and form some of the most productive and stable agricultural areas in Latin America.

Environmental sensitivity and change: the long-term impacts

The environment is innately dynamic and human activity tends to accelerate or retard natural processes. In this section, the longer-term consequences of vegetation change, land degradation and coastline change in Latin America will be examined – as examples of the regional and global significance of environmental disturbance.

Tropical deforestation

The most publicized environmental changes are those of tropical deforestation and soil erosion. Forest resources not only affect the traditional Amerindian and *caboclo* inhabitants but many of the small-holders, whose poverty drives them to marginal environments. The broad problem is well known with the ecological arguments for protecting forests including their unparalleled biodiversity, their role in environmental maintenance and protection and their potential as well as present uses, set against the forces of economic exploitation, poverty, landlessness and personal greed. The actual situation at any one place is, however, complicated by numerous other factors. They include the type of forest (Table 4.2) which affects its attraction for developers and resilience against development, the evolution of the forest (for instance whether it is on the margins, therefore vulnerable to change), the proximity to the agricultural frontier or other form of development and the nature and intensity of disturbance. The most endangered forests are not, as often thought, the tropical moist forests, but the lesser known and more heavily damaged systems such as the Atlantic forest of Brazil which now only exists in remnants, or the tropical deciduous forests which are often located over mesotrophic soils and consequently have been used over a long period by local people.

Table 4.2 Classification of tropical forests

Tropical everygreen forests:
Rainforests composed mainly of
evergreens:

1. *Lowland forest*: multilayered
structure and many trees exceeding
30 m in height.

2. *Mountain forest*: tree sizes
markedly reduced and few species
exceed 30 m; abundant undergrowth.

3. *Cloud forest*: closed structure with
numerous gaps and liana thickets;
trees often gnarled, rarely over 20 m.

4. *Alluvial forest*: multilayered closed
structure with numerous gaps,
distinctive herbaceous undergrowth
and palms; many buttresses and stilt
roots.

5. *Swamp forest*: tree species limited
and usually less than 30 m high; stilt
roots and air-breathing roots common.

6. *Peat forest*: stands generally less
than 20 m high with a few, slow-
growing, broad-leaved trees or palms.

**Tropical and sub-tropical seasonal
forests**: dominantly evergreen and
leaf-exchanging trees. Upper canopy
species reach around 40 m and are
moderately drought-resistant.
Transitional lowland, montane and
dry sub-alpine forests.

**Tropical or sub-tropical semi-
deciduous forests**: most upper layer
trees are drought resistant and shed
leaves fairly regularly in the dry
season. Some emergents reach 40 m.

1. *Lowland forest*: multilayered
structure with marked emergents
(some 30 m); undergrowth of tree
seedlings, saplings and woody shrubs.
2. *Montane forest*: similar structure to
other montane and some cloud forests
but with smaller trees.

Sub-tropical evergreen forests:
multilayered stands with similar
divisions to tropical evergreen forests
but less vigorous and with more
shrubs in the understory.

Mangrove forests: Halophyllous
(salt-adapted) forests of intertidal
zones in the tropics and sub-tropics;
single layer trees up to 30 m; various
kinds of air-breathing adaptations.

Projections from current rates of deforestation give estimates for the survival of tropical forests varying from less than 30 to 100 years or more. Whichever projection or set of statistics is favoured, the signal is the same – that at the present rates of depletion, reafforestation and conservation, there will be little left to pass on to generations two or three removed from the present.

The most recent comparative figures (FAO 1992, 1993) subdivide tropical forests into ecological categories (Box 4.6). The overall impression is that areas affected by the customary definition of deforestation (that is clear felling often accompanied by burning) are being exceeded by the area of forest fragmentation and degradation. This means that the depletion of habitats is occurring much more rapidly than the deforestation figures alone would imply. Deforestation occurs when crown cover is depleted to less than 10 per cent of its original – characterized by distinctive wild flora and fauna, natural soil conditions and no agriculture. For the first time, data are available for specific forest ecosystems, showing the levels of degradation and changes in biomass (around 50 per cent of the global biomass loss of 2.5 gigatonnes a year occurs in Latin America). The FAO figures, however, require cautious appraisal since the methods of data collecting, the comprehensiveness of data and data reliability vary widely between different national reporting centres.

Box 4.6: Tropical deforestation

The published figures make depressing reading for Latin America. Brazil is in the process of losing the greatest area in virtually all categories of forest – montane, dry, moist deciduous and moist evergreen (Fig. 4.12). Throughout tropical America in the past, the greatest forest losses have been in easily accessible lowlands. In 1993 the FAO estimated the losses to be 1.9 million ha y^{-1}, but the pace and spread of disturbance as well as land cover change, has now moved into the marginal and remote areas. Latin America lost 3.1 million ha y^{-1} of moist deciduous forest and Brazil alone lost 310,000 ha of dry deciduous forest, whilst Mexico and Brazil have the highest annual losses of the unique montane forest.

There are still very large forest reserves and it is the rate of depletion rather than simply the area lost that is of concern. In 1990, Brazil possessed over 561 million ha of tropical forest with a 0.6 per cent annual change in the period 1981–90. Of this figure, nearly 292 million ha was made up of rainforests (0.3 per cent annual changes since 1981), 197 million ha consisted of the endangered moist deciduous forest (0.8 per cent change), 29 million ha is dry deciduous forests (1.0 per cent change) and 44 million ha was in hill and montane forest (1.3 per cent change 1981–90).

Many of the forest tracts suffer from degradation and fragmentation, which may be more important than the outright conversion of forests to other forms of land use. Degradation results in the selective removal of trees and other plants which may be essential to the life strategies of other species, and increases the risks of erosion. Similarly, fragmentation can result in a patchwork of forest for which interconnectivity has been lost or weakened. Such fragments are vulnerable to drying and 'edge effects' as well as fire, and the total aggregated area of forest cannot support the populations of plant species which might be estimated from area alone. A NASA study of 1978–88 showed that only 39 per cent of the altered forest habitat in the Brazilian Amazon Basin came from outright conversion, whilst 58 per cent came through exposure to newly cleared areas at forest edges from microclimatic changes, human disturbance and invading species. Deforestation is moving from the edges of forest reserves to the centre, criss-crossing the areas with roads, cultivation and ranching.

Roraima in the northern Brazilian Amazon, provides a good illustration of the pace of events in an area remote from the industrial south. In a survey of LANDSAT TM and MSS satellite images between 1978 and 1985 for the northern part of the State close to the Venezuelan border, it was found that around 10 per cent of the total forest area had been converted to other uses. More than 4 per cent had gone in the remote western area and over 30 per cent in the developed east, around the capital Boa Vista. This high figure reflects the loss of the irreplaceable *várzea* (riverine) forest.

Sources: WRI (1994); Uhl and Kauffman (1990); Skole and Tucker (1993); Dargie and Furley (1994).

Soil erosion and land degradation

Although the seriousness of erosion has been known for centuries, it has moved to the forefront of world attention again in the past 20 years. Indeed in the Report to the President, Global 2000, the economist editor commented that soil erosion was the 'World's No. 1 problem'. Its frequently imperceptible character, contrasting with the much more spectacular, but highly localized, gulley erosion, tends to camouflage the long-term degradation of resources. Since erosion – the 'gnawing away of land' – works mainly down from the surface, so the character of the surface soil (the organic and frequently nutrient-rich rhizosphere) is the horizon most affected (Box 4.7). Soil is not an inert mass but a highly dynamic and delicately balanced assemblage of mineral, gaseous, aquatic, organic and living organisms. The key properties of rooting depth, structure (bulk density, permeability and stability) and physical chemical attributes (degree of acidity, nutrient ions and availability, oxygen and water availability) are particularly sensitive to disturbance and change in the surface horizons. Equally, because the soils of the humid tropics and temperate areas are frequently maintained by intense nutrient cycling, so the destruction of the surface vegetation breaks the regeneration process and leads to a weaker structure and resource base for plant growth (Fig. 4.12).

Box 4.7   Impact of land clearing methods on soil properties

The method of clearance as well as the act of clearance itself has been shown to be important in the sustainable use of land. Land clearance is the most crucial step affecting future productivity of farming systems.

The physical removal of vegetation takes many forms, from small-scale manual cutting of selected trees with little or no burning, to large-scale mechanized clear felling, bulldozing and combustion of the entire soil surface. As a generalization, the greater the intensity of the clearing process and the larger the scale of operation, the more long term are the environmental problems left for future management.

Manual cutting is one of the most common methods for the poorer smallholders and illegal shifting cultivators since it requires the least capital expenditure. If the fallen litter is then burnt, there is an immediate beneficial effect for food crops resulting from the fertility and nutrient availablity in the ash and the temporary buffering of excess acidity (for example from aluminium levels). At the same time the nutrient store of the original vegetation has now been plundered and rapidly disappears in smoke and gaseous diffusion (especially N and S) or is leached or washed away in subsequent rains (in particular base cations). Increasing the level of mechanization increases the compaction of the soil and is likely to affect permeability and aeration and, therefore, regeneration (Fig. 4.13).

Box 4.7   Impact of land clearing methods on soil properties
(continued)

Effect of land clearing methods:
(a) On topsoil physical properties and organic carbon on a continuously cropped Ultisol in Yurimaguas, Peru.

Clearing method	Months after clearing	Infiltration rate (mm h^{-1})	Bulk density (mg m^{-3})	Mean weight diameter (mm)	Organic C (%)
Before clearing	0	324	1.16	0.48	1.04
Slash and burn	3	204	1.27	0.42	1.05
Straight blade	3	14	1.42	0.29	0.82
Shear blade	3	22	1.28	0.36	0.87
Slash and burn	23	107	1.32	0.38	1.03
Straight blade	23	15	1.42	0.36	1.02
Shear blade and disk	23	110	1.32	0.36	0.89

(b) Including post-clearing management, on the relative yield of five consecutive crops after clearing a 20-year-old secondary fallow on an Ultisol of Yurimaguas, Peru.

Clearing methods	Percentage of cumulative maximum yields	
	No tillage, no fertilizer	Tilled, fertilizer
Slash and burn	27	93
Straight blade	7	47
Shear blade	14	65
Shear blade + burning + heavy disk	28	89

Note:
1. Maximum yields of five consecutive crops in t ha^{-1}: upland rice, 4.0; soybean, 2.3; maize, 5.2; upland rice, 2.5; maize, 3.3
Source: Furley (1990) based on Lal *et al.* 1986.

According to the WRI *World Resources Report* (1992/93) based on a UNEP (1990) *Global Assessment of Soil Degradation,* and the *World Map of Soil Degradation* (Oldeman *et al.* 1991), around 1.2 billion hectares of the Earth's surface has been moderately, severely or extremely degraded since 1945. To give an idea of the scale of this figure, it exceeds the area of India and China combined, and of this sum, around 300 million hectares have lost all original biological productivity.

A greater percentage of land has been degraded in Latin America since the Second World War than anywhere else in the world (Figs. 4.13

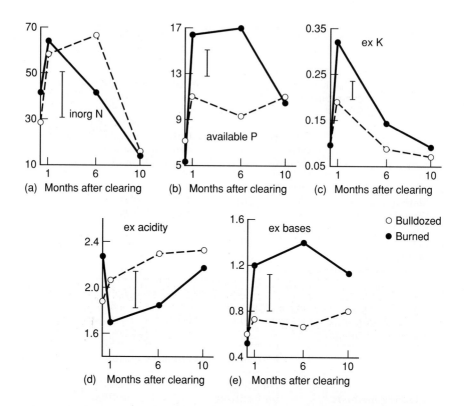

Fig. 4.12 Effects of vegetation clearance on soil properties. The major nutrients (N, P, K) are frequently released after clearance and burning, giving temporary benefit for plant growth (as shown in the exchangeable bases). At the same time, excessive acidity, largely Al-based, is buffered and its impact diminishes over the first year. Subsequently, nutrients tend to revert to their normal low levels but with clearance of vegetation the soil is now bereft of its organic input. If soils are not given time to regain their pre-clearance level, then further disturbance depresses nutrient availability even more. (*Source*: after Seubert 1977)

and 4.14). The environmental consequences form a well-known litany: losses caused by soil erosion and lowered fertility loss, desiccation through to extreme desertification, deforestation and vegetation loss, depletion of all other biological resources, reduced biodiversity, accentuated run-off often leading to flooding, depletion of groundwater, impacts on aquatic life and pollution. In dry areas, wind erosion becomes more important, sifting away the fine and often more fertile particle sizes at wind speeds over around 20 kilometres per hour. All these processes combine to reduce carrying capacity, land productivity and biological usefulness and have positive feedbacks augmenting the

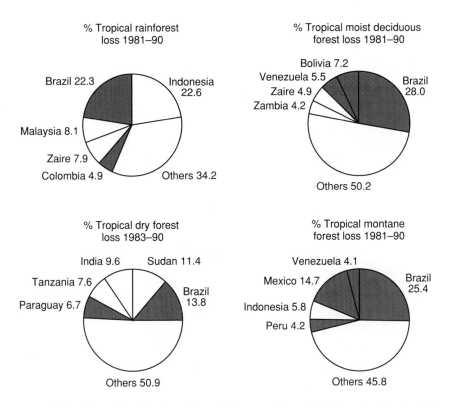

Fig. 4.13 The pace of tropical deforestation: percentage forest loss
(*Source*: after FAO 1993)

problem. At best, regeneration of degraded areas requires increasing inputs to maintain the same level of yield.

Land degradation (LD) at any given site can therefore be defined under its main component categories (modified from FAO/UNEP 1983):

LD =   vegetation and organic depletion + desertification
        + erosion + salinization + toxic accumulation
        + waterlogging or drainage modification.

One of the first indicators of decline is the loss of organic matter, which may drop from somewhere in the order of 2–7 to below 0.5 per cent. With the loss of the protective vegetation and organic matter, the soil becomes vulnerable to rainfall impact and the sheer volume of water arriving at the surface (50 millimetres of rain = 50,000 tonnes of water every square kilometre).

Land degradation is a particular affliction in poor countries. The World Bank has a definition of low income countries (1991) of an

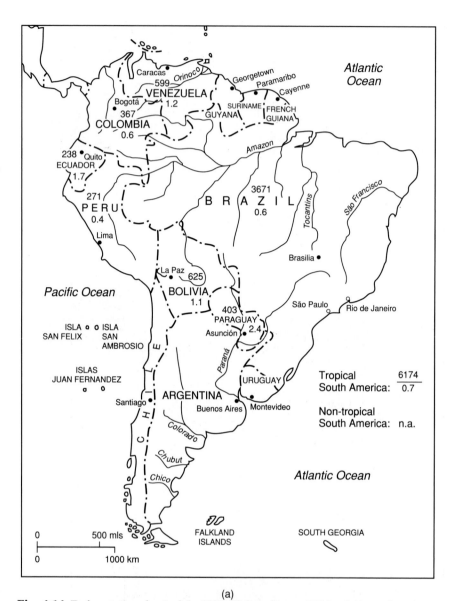

(a)

Fig. 4.14 Deforestation: basic data 1981–90 (top figure, 000 ha; bottom figure, percentage): (a) South America; (b) Central America and the Caribbean (*Source*: data from FAO 1993)

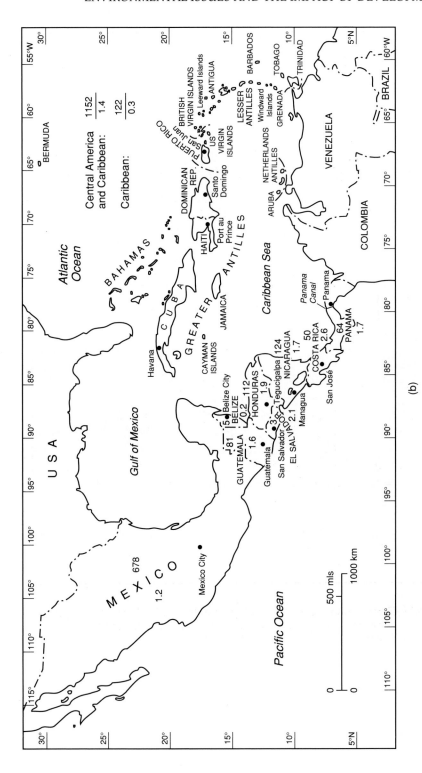

Fig. 4.14b

average annual per capita gross national product (GNP) less than $580 (1989 figures). In Latin America, Central America and the Caribbean, only Haiti and Guyana fall into this category, but there are many regions within other states which show comparable signs. Poverty has been shown to be closely related to environmental degradation: 'The evidence of human poverty and depravation in the world is unmistakable, as is the evidence of worsening environmental conditions caused by and contributing to poverty' (WRI 1992: 30).

As a consequence of rapid population growth, modernization of agriculture and inequality of land tenure, larger numbers of people have little or no access to productive land and are driven to marginal environments. These are frequently unsuited to sustaining a food base and yet represent the agricultural frontier in many parts of Latin America. It has been suggested that around 60 per cent of the developing world's poorest people live in areas that are ecologically vulnerable.

Rapidly industrializing countries have additional problems as well as benefits. This group includes Brazil, Chile and Mexico. Industrialization

Plate 4.1   Building part of the Carretera Marginal de la Selva in Peru. The Carretera Marginal was the idea of former President Belaúnde of Peru and was designed to run along the eastern margin of the Andes. Some parts have been completed but the demand for access is east–west rather than north–south. Clearing forests on steep slopes causes considerable erosion. (Photo: Tony Morrison, South American Pictures)

widens the scope of land degradation and includes ground as well as air and water pollution (see the section on Problems of the built environment).

Coastal environments

Coastlines are particularly vulnerable to damage since they tend to be the focus of concentrated human activity. Much of the colonial and post-colonial development of Latin America has taken place from landings on the coast followed by penetration along rivers or other physically controlled routes to the interior. It is not surprising therefore, that the coastal inshore waters and the littoral vegetation have been subject to the most pressure. Nowhere is this more evident than in the Atlantic Forests, stretching from Vera Cruz in Mexico to south of Rio de Janeiro in Brazil.

More recently, coastal environments that were previously protected by remoteness or their inhospitable nature have also experienced incipient development. Such environments include the arid deserts of Peru and northern Chile, the mangrove swamps of the Atlantic side of Central America from the Yucatán to south of the Amazon estuary, other wetlands and lakes such as the Pantanal between Bolivia and Brazil, and the coral reefs of the Caribbean and Belize–Mexico. The main impacts have been commercial and industrial – increasing inshore and coastal fishing, developing ports and trading links, exploitation of timber and game and, most recent of all, the exploding tourist industry with its quasi-environmental face of eco-tourism (Box 4.8).

The impact of development schemes on the environment

Any disturbance to the natural, dynamic equilibrium of the environment involves some element of damage. In some instances the effects of disturbance are only felt after decades or even hundreds of years, such as the gradual depletion of soil nutrient resources. Other impacts are much more immediate with severe, often dramatic effects on the environment, such as hydroelectric power (HEP) projects. The critical question is whether there are thresholds beyond which the environment can no longer sustain the developments to which it is being subjected. The long-term answers to this question are frequently not yet apparent, but the short-term effects of specific projects have been explored in some detail.

In this section, a number of these projects are examined to reflect the range of developments currently problematic in Latin America. On the one hand, the impacts of mining, the oil and HEP industries are relatively localized and their effects can be measured with some accuracy. On the other hand, some developments have a more disseminated influence such as the processes of land colonization; their

Box 4.8 Mangrove coasts and coral reefs: the case of Belize

Coastal environments have proved to be particularly sensitive to disturbance. Mangrove and coral provide two examples, forming linked ecosystems which share common problems for very different reasons. The Barrier Reef lying to the east of the Yucatán peninsula is of special interest since it is by far the largest in Latin America and second only to the Australian Barrier Reef in the world.

The Belizean coral reefs and atolls extend for about 100 km out from the shore and run some 325 km down the length of the country. The largest, longest-settled and most affected cays (coral islands) lie to the north, whilst relatively undisturbed corals are found on the atolls and in the south. At present, the coral reefs are threatened from several sources: increasing sediment and pollution from the mainland, along with sharply increased tourism, fishing and diving from the cays. There is concern that the carrying capacity of the cays such as Ambergris or Cay Caulker (Fig. 4.8) could be rapidly exceeded if the number of visitors continues to rise without restraint: the damage to the coral from physical contact and pollution could quickly degrade the very resource that visitors have come to see. Surveys and assessments of thresholds are currently being undertaken to furnish evidence for the coastal zone management plan.

The threat to the mangroves comes from a different direction. Mangroves have traditionally been perceived as unpleasant obstacles to development, except by the few fishing communities who have utilized the mangrove resources. Mangroves have been considered as valueless in the past and this has led to indiscriminate clearance for housing (about 70 per cent), tourism (15 per cent), industry (5 per cent), transport (5 per cent) (ports airstrips, yards), waste disposal and aquaculture (10 per cent). Most of this development has been around the largest settlement of Belize City. Recently, the considerable economic and recreational value of mangrove has become apparent, especially its role as a nursery for restocking the brilliant and diverse fauna of the related cays, its value as a rich habitat for wildlife, especially birds, its protective role against storms and tidal surges, as well as its function as a sediment trap for suspended particles and pollution brought down by rivers and currents.

They are systems in delicate equilibrium, where relatively small-scale developments or accidents such as oil spills could create devastating damage to the environment.

Sources: Zisman (1992); Furley and Ratter (1992).

ultimate impression on the environment may be greater but they are much less amenable to prediction.

Mining impacts

Mining has been a characteristic feature of life in Latin America since pre-Conquest times. At first, the extraction of gold, silver and diamonds was extremely localized with relatively slight impact. Damage to the environment has grown steadily since then, with an increasing range of mineral discoveries. The advent of satellite and other remote sensing techniques (for example Projeto Radambrasil) has greatly increased the range and scale of activity. Mining impacts can be divided into two: the informal economy (*garimpagem*), which is small scale and often unofficial with scattered but cumulatively important consequences; and the formal sector, represented by the large commercial, national and international companies, which can devastate very large stretches of the environment, particularly where the operations are opencast with an absence of legislative controls.

The main metal ores are found in the Andean cordillera, Precambrian shields and Mexican plateau. River basins tend to accumulate ores in placer deposits and these have also been extremely vulnerable environmental sites. Amongst the main minerals which have been mined and have affected the environment are copper, tin, iron and gold. Other minerals which have had an important extractive effect are bauxite (Brazil, Suriname and Guyana), lead (Peru and Mexico), mercury (Mexico), nickel (Cuba) and zinc (Peru and Mexico).

Gold is one of the most widespread mining activities, particularly the small *garimpos* found scattered over numerous rivers and streams. The mining has the effect of churning up sediments, increasing the suspended load and damaging the aquatic food chain. A related impact is mercury pollution, since mercury is used to amalgamate the fine gold particles in the extraction process. An estimated 250 tonnes were dumped into Amazonian rivers in the four years between 1984 and 1988 (quoted in Fearnside 1990). Since mercury is extremely toxic, its uncontrolled use increases the health risks to the operators and to all those who eat contaminated fish.

Copper is one of the most important ores with a large international trade. Chile and Peru are the principal producers, with Chile being the world's largest producer in 1992. Since copper is only present in very small quantities, the extractive process is very large scale, creating physical disturbance and also generating dust. The process of concentration results in toxicities in the surrounding soils and vegetation and also in downwind or downstream pollution. However, the Latin American reserves of copper are limited and this may affect the future scale and intensity of operations.

Tin is another mineral where the mining industry has had a major environmental impact. The growth of the industry in Brazil has been spectacular with large commercial operations opening up in the Amazon Basin (Rondônia and Roraima), often threatening Indian settlements and reserves as well as causing major open-cast physical damage and the same polluting processes as those described for copper. Bolivian production halved between 1977 and 1992 despite remaining the cornerstone of the economy, and the reserves are estimated to be significantly less than those of Brazil. In some places, tin is mined by *garimpagem* alongside gold, although the refining and concentrating process is not so suited to small-scale operators.

Iron mining has had a less boom and bust type of development but is another mineral with a widespread role in environmental damage. The extraction tends to be large scale, often with international trade outlets which, like so many of the mineral industries, have dominated and directed the course of development. Much of the low-grade ore, for example Sierra Grande in Argentinian Patagonia, requires very large-scale excavating operations, as witnessed over many years in the rich metalliferous area around Belo Horizonte in Brazil. Many of the ores are high grade, but the scale of mining with its associated smelting, transporting and exporting components has created a marked impact on the environment. Grande Carajás, the largest iron ore operation in the world, exemplifies this type of development and its influence on the surrounding landscape (Box 4.9).

Fuel and energy impacts

The quest for energy sources has traditionally tended to ignore the environment and, for that matter, the local people. Much more care has been evinced over recent years and nearly all developments require Environmental Impact Assessments (EIA). However, these are often superficial and seem at times to be simply gestures to placate the funding agencies. Most World Bank and other investment loans are heavily hedged about with phrases designed to protect the environment; despite this, one of the most frequent criticisms of development projects is their lack of concern for or protection against long-term environmental degradation.

Nevertheless the impact of these essential industries on the environment has been prodigious, if unevenly distributed. Coal has never been a major extractive industry in Latin America, although localized mining activities are found in Brazil, Chile, Colombia and Mexico (the last two having the largest reserves of high-grade fuel). As an overall generalization, the coal industry is not in the same league of physical damage or pollution effects compared with the industries of Eurasia or North America. Crude oil extraction is an extremely important industry in Mexico and Venezuela (with some of the world's largest reserves), and to a lesser extent in Argentina, Brazil, Colombia and Ecuador. Considering both trade and distribution as well as extraction, the oil industry has made a major impact on the environment. The concerns of environmentalists reflect the damage to the biology of an area following oil spills, air and water pollution from the petrochemical industry, as well as the physical impact of the widespread and large-scale infrastructure required. Oil industry developments and their possible environmental effects are highly contentious. Oil is essential, and petrol and diesel engines power most of the transportation networks for the region. Consequently the clash between environmentalists seeking to protect landscapes, wildlife and indigenous groups and governments/developers concerned with the exploitation of resources, has witnessed some of the most bitter clashes of all the extractive industries (see for example the Amazon region of Ecuador) (ECLAC 1989).

The development of HEP has had an equally controversial history. Brazil has by far the largest installed capacity in Latin America and, at least theoretically, an enormous potential. Mexico, Venezuela and the Andean countries also have a large existing capacity and potential – so the direct impact of dam construction and the indirect effects in and

Box 4.9 Grande Carajás

The Greater Carajás Programme (PGC) is the largest development project of any tropical forest area, affecting around 900,000 square kilometres or over 10 per cent of Brazil. It opened in 1980 and is based on the rich iron ore deposits discovered during Projeto Radambrasil. The heart of the programme is the extraction, processing and export of minerals but there are also associated forestry, agricultural and ranching activities.

The PGC has sucked in very large capital investments from within Brazil and from abroad and drawn immigrants from the whole of Brazil. The populations of local towns such as Marabá, Açailândia and Imperatriz have mushroomed. The main components of the programme are iron ore extraction and smelting (with manganese, gold, nickel and tin), two related aluminium plants at Belém and São Luis, the Tucurui HEP scheme and the 900 kilometre railway line from the ore field to the Atlantic at the new deep water port at São Luis. There were a number of motives behind the grandiose nature of the scheme, which reflect many of the problems affecting Latin American development. The Brazilian government had already incurred a very large foreign debt through projects like the Trans-Amazonian Highway and the Iguaçu HEP Scheme. The Amazon road development had not generated the expected export earnings. There was a growing state interest in agri-business and mining with the potential of the latter revealed during the Radam surveys. Expansion in the 1970s was achieved through the second National Development Plan (1975–79) and the Polamazônia proposals, which were a series of development nucleii throughout Amazônia Legal. In addition to US involvement from the beginning and Japanese capital investment at a critical time in the early 1980s, other foreign investments included a major EC contribution. The Japanese influence, in particular, helped the diversification into agriculture and timber extraction but all of the projects have been criticized on the grounds of increasing national debt, increasing dependency relationships, for their effects on the environment and for not addressing the interests of local inhabitants.

The total effect of these projects upon the environment has been enormous. The Tucurui Dam sparked off intense controversy. The dam, located 300 kilometres south of Belém, backs up water for 200 kilometres along the Tocantins river and floods 2,160 square kilometres of forest. This area could not be completed cleared and led to anaerobic decay of the trapped organic matter with the production of methane and hydrosulphide (see Box 4.10). A number of other major impacts have been identified, from the destruction of the Brazil nut habitat and local fish resources, to increases in water-borne diseases, land degradation over the catchment and profound effects on the Indian communities. These have accentuated the pre-existing social tensions in the area by forcibly resettling around 15,000 people and encroaching on six indigenous reserves. Although a number of steps have been taken to monitor the environmental effects, including the formation of an independent advisory group, the continued deforestation, the widespread land degradation and relative lack of restraint and control have caused international concern.

Carajás provides a classic example of relatively short-term planning which has not taken the long-term environmental costs into anything like sufficient account.

Sources: Monosowski 1991; Hall (1989)

around HEP sites have drawn much adverse environmental comment (Table 4.3).

Table 4.3: Installed capacities and known exploitable potentials for hydro-electric power (1990) (megawatts)

	Installed capacity * (over 500 megawatt installed capacity)	Known exploitable potential † (megawatts)
Central America		
Costa Rica	754	37
Mexico	7,837	80
Panama	551	16
South America		
Argentina	6,499	390
Brazil	45,558	1,117
Chile	2,290	162
Colombia	7,201	418
Ecuador	897	180
Paraguay	5,790	40
Peru	2,396	412
Uruguay	1,196	7
Venezuela	7,914	262

* Many countries which have little current installed capacity, have significant potential, such as the Central American countries lying along the Cordilleras and the Andean countries plus Guyana
† Much of the known 'exploitable' potential would cover areas greatly disputed by environmentalists

The future projects within Brazilian Amazonia are outlined in Plan 2010. Since Brazil has insufficient supplies of fossil fuels, HEP is seen as the principal source of energy, currently supplying 95 per cent of the country, and the major development potential has concentrated on Amazonian rivers. Long-distance transmission lines are planned for the industrial south-east of the country. Plan 2010 has provision for 31 HEP stations but double that number have been considered and many are planned for the highly disputed Xingu River. A number of environmental protection guidelines have been issued but the costs and benefits of these schemes are not felt equally. The negative effects are felt at source, in the Amazon, whereas most of the benefits are felt outside the region.

Brazil has numerous very large HEP schemes. The giant Iguaçu scheme on the borders of Argentina and Paraguay illustrates many of the problems and arguably successes of large-scale schemes. The economic advantages have to be set against environmental costs however, and few development projects have ever considered restoration costs along the lines being advocated by a number of environmental economists. At Tucurui for example, little emphasis was given to the interests of the indigenous people or to the special character of the natural resources, and practically no attempt was made to value the socioeconomic activities of the region. A recent, equally controversial, project has been designed to supply the fast-growing city of Manaus (Box 4.10).

Box 4.10: The Balbina dam – HEP Project

The Balbina dam is a more recent example of the controversy surrounding large-scale projects. Balbina lies on the Uatumá River some 150 kilometres north of Manaus which it was built to serve. Construction began in 1981 and the HEP station opened in 1987. The flooded area is around 2346 km^2, close to that of Tucurui, but the capacity is much less (250 MW installed capacity of which only a third will be generated, as opposed to 9000 MW at Tucurui), resulting from the flat topography and the area required to generate sufficient capacity (Gribel 1990). Part of the area inundated by the reservoir had only been designated as within the Waimiri-Atroari Indian Reserve in 1971. The cost, 750 million US dollars excluding transmission lines is considered to be the most expensive, yet poorest designed, of Brazilian HEP schemes.

The traditional lands of the Waimiri-Atroari are believed to cover an area of 8 million ha. The reserve created in 1971 covered 1.146 million ha, whilst the population has declined from an estimated six thousand at the turn of the century to less than 500 in the mid-1980s, mainly as a result of contact with outsiders. A third of this number was resettled as a result of the dam and reservoir construction. Despite the designation of reserve status, the area was first affected by the BR174 (Manaus–Boa Vista–Venezuela road) which caused violent clashes, and later by the mining operations of the very large Paranapenema Company.

The main problems and consequences of the Balbina HEP project can be summarized as follows:

Problems:
1. There were insufficient initial surveys to give an adequate picture of the natural resources or of the inhabitants of the region.
2. Inadequate attention was given to the downstream and regional impacts of the programme.
3. The estimates for the amount of energy produced and the area to be flooded were inaccurate.

Consequences:
1. It has been a major disaster for the Indian communities, with disease, resettlement and disturbance to their way of life.
2. The project design is widely considered as unsatisfactory, flooding a very large area and still producing insufficient energy to supply the Manaus area.
3. The size of the flooded area has had a proportionate effect on the wildlife.
4. Little of the forest could be removed before filling, leading to anaerobic decomposition of the biomass, increasing acidity and seriously disrupting aquatic life.
5. There has been an increase in disease particularly affecting the Indian communities, notably of malaria and leishmaniasis, and this has spread to other areas such as the town of President Figueredo.
6. The downstream communities have also been affected by changes to the water supply, fishing and *várzea* (flood plain) cultivation.
7. The cost of generated power has been very high and compares unfavourably with other Amazonian schemes (producing 25 less power per square kilometre than Tucurui and 35 times less than Itaipu).
8. The whole reservoir may arguably be shown to have been unnecessary by the construction of Cachoeira Porteira dam.
9. The Balbina dam construction has led to a surge of criticism and discontent, rivalling or even exceeding better known projects elsewhere.

Sources: Cummings (1990); Goldsmith and Hildyard (1984), (1986); Trussell (1992).

Large-scale projects such as these have generated fierce arguments between developers and conservationists. The understandable need to exploit resources can be set against the frequently unsympathetic and short-sighted policies used to achieve development. The wishes and interests of local people are rarely consulted, whereas foreign companies and institutions, distant from the site, are perceived as reaping most of the benefits. Environmental concerns have achieved a high political profile, as witnessed by the UNCED meeting (the United Nations Conference on Environment and Development) of 1992, but, in practice, there is often bitter local opposition to any controls which appear to restrain economic progress. What is now evident, is that the massive scale of this development is having a regional and at times global effect, and therefore concerns the world community. The local power of national pressure groups including non-governmental organizations (NGOs), as well as international lobbying, is providing a brake on environmental degradation but it has not stopped or limited the momentum of development.

Problems of the built environment

The scale and intensity of environmental change are nowhere more keenly felt than in urban areas. The proportion of South America's population living in urban areas has risen from 56 per cent in 1965 to 78 per cent in 1995, and from 42 to 56 per cent in Central America and the Caribbean (WRI 1994). In the more developed countries, the proportions are even more extreme: for instance Mexico 55–75 per cent, Brazil 50–79 per cent, Argentina 76–88 per cent and Venezuela 70–93 per cent. Already Latin America is as urbanized as Europe. The impacts are seen not only directly as a result of transformed physical structures but indirectly in the atmosphere, in water quality and on adjacent land resources. They mirror the prodigious rises in urban populations.

Urban air pollution

Urban air pollution is perhaps the most striking environmental impact. The world's mega-cities share similar problems and are growing at unprecedented rates. By the year 2000, they may account for over 50 per cent of the world's population. It has been estimated (WHO/UNEP 1992) that by 2000 there will be 85 world cities with 3 million or more inhabitants and 11 of these will be found in Latin America. They are usually but not always state capitals: Mexico City, São Paulo, Buenos Aires, Rio de Janeiro, Lima-Callao, Bogotá, Santiago, Caracas, Belo Horizonte, Porto Alegre and Guadalajara. At the same time, many of the urban inhabitants of Latin America, as elsewhere, have poor shelter, a lack of safe drinking water and poor sanitation, waste management and drainage, giving rise to health risks which are accentuated by harmful levels of air pollutants from industry, vehicles and domestic combustion. Latin American cities tend to have higher vehicle densities

Plate 4.2 Smog and sunset: a view across central Mexico City. (Photo: Kimball Morrison, South American Pictures)

than other developing countries and therefore higher vehicular contribution to air pollution.

An increasing number of toxic and carcinogenic chemicals have been detected in these large urban agglomerations. They include heavy metals (such as cadmium or mercury), trace organics (benzene, formaldehyde, vinylchloride, polychlorodibenzodioxins and polyaromatic hydrocarbons), radionuclides and fibrous wastes. They are released from vehicles, industrial and manufacturing processes, waste incinerators, sewage treatment plants and building sites.

The dispersal of pollutants is highly specific to each urban area though clearly, similar patterns of heat islands, input processes and outputs prevail in all mega-cities. The pollution effects are aggravated by local meteorological and topographic factors, which may concentrate them or retard dispersal and dilution.

Even amongst these mega-cities, São Paulo and Mexico City stand out as giants. They are not only amongst the largest urban areas in the world but also have giant problems (Box 4.11). São Paulo for example, contains 10 per cent of the Brazilian population spread over 8,000 square kilometres and had over 4 million vehicles in 1990. Greater São Paulo has witnessed a slight decline in indicator gases such as sulphur dioxide since the early 1970s, as efforts have been made to reduce pollution. However, the proportion of suspended particles in the atmosphere is still serious, there are high smoke levels and, although

---

Box 4.11: Urban pollution in Mexico City

Mexico City has endured one of the most dramatic rises in population anywhere in the world, achieving the dubious distinction of having the largest number of inhabitants, estimated at 24 million. Spread over 2,500 square kilometres, at an altitude of 2240 metres, its plateau location in a shallow basin has accentuated the accumulation of toxicants and the city has had an unenviable reputation for pollution since the early 1970s.

The sources of air pollution are industry, power stations and vehicle emissions. There are an estimated 36,000 light and heavy industrial plants which, together with the power stations, contribute 30 per cent to the pollution load over the Central Valley. However, the major polluter is the motor vehicle, 3 million of which contribute 40 per cent of the total pollution. In all, some 5.5 million tonnes of contaminants are released into the air every year.

Several indicators highlight the seriousness of the pollution. For example, the sulphur dioxide levels have been well over World Health Organization (WHO) minimum figures for 20 years. There are high carbon monoxide, nitrous oxide and ozone levels. The latter help to generate a photochemical smog, particularly in the south-west, channelled by the north-east winds which blow over the city. In addition, the clearance of vegetation, with expansion of the urban area, and the drying up of the old lake bed have produced soil erosion which generates high levels of suspended dust, coating the city in a grey blanket. Inadequately controlled waste dumps have also provided a health hazard, scattering pollution over the urban areas in high winds.

Not surprisingly, the effect of pollution on the health of the city's inhabitants has been growing. Some 2 million people are believed to be suffering from diseases directly resulting from or aggravated by air pollution. In March 1992, schoolchildren under 14 were asked to stay at home for a month as a result of high dust concentrations and staggered school hours have been introduced in places.

Steps have been taken since 1990 to reduce pollution levels. Various international loans have been negotiated for a city clean-up. Lead levels have dropped slightly as unleaded fuels are used for vehicles and restrictions on car use have had some impact. Catalytic converters on buses and taxis and replacement of older vehicles have also made a limited contribution. Despite this, there is a continuous increase in petrol consumption. However, in 1991, the PEMEX state oil refinery was closed – which removed a major source of pollutant gases. City services have had an impossible task in keeping up with population increases and maintenance defects have been blamed for an increasing number of incidents, such as the gas explosion of 1983.

Sources: Gilbert (1994), WHO/UNEP (1992), WRI (1992), Ward (1990)

---

lead levels from motor exhausts are less with the greater use of alcohol-based fuel, the scale of carbon monoxide, nitrogen dioxide and ozone contaminants is very high (WHO/UNEP 1992). Steps to control car use have also been taken in Caracas, Santiago and Mexico City, but petrol consumption is still increasing rapidly (for example, by 18 per cent from 1988 to 1991 in Mexico City; *The Economist* 4 April 1992).

Industrial areas have provoked numerous environmental problems and little control has been exercised over large manufacturing companies. A well-known example is at Cubatão on the Atlantic close to Santos in Brazil, which had 24 large steel, petrochemical and cement companies operating in the early 1980s. Cubatão alone is estimated to have produced 3 per cent of Brazil's GNP in 1985 (WRI 1991). These combined to push over 200 tonnes of particulate matter per day into the atmosphere (Gilbert 1994). Since the city lies in a topographic depression at the foot of mountains and is liable to temperature inversions, it trapped this contaminated air. The smog affected plant growth which, together with forest clearance led to severe hillside erosion and high sediment problems in the rivers. In the mid-1980s conditions became so bad that a state of emergency was declared and parts of the city of 110,000 people were evacuated. A later atmospheric clean-up reduced particle emission by nearly 90 per cent, with considerable reductions in ammonia, hydrocarbons and sulphur but less successful nitrogen oxides. Water quality has also improved in the rivers, but this degree of success is not typical of most Latin American industrial complexes. Air pollution has reached the stage when it noticeably affects plant growth in cities.

Domestic waste has long caused environmental problems in and around major urban areas. Burning waste and water pollution also increases the risk of pests and diseases and untreated sewage carries similar risks. For example, Santiago pumps around 300 million cubic metres of sewage into two rivers and a canal flowing through the city and is believed to be partly responsible for the high rates of typhoid. Greater risks face the poorest sectors of the population where environmental controls are frequently non-existent. The *favelas* of Brazil or *barrios* of Spanish-speaking Latin America represent the margins of life in some of the world's most degraded environments.

Two further situations may be considered briefly, where construction has had a major impact on the environment. These concern the pressures caused by the road-building programme and the impact of government, political and strategic policies.

Road building

Latin America is relatively undeveloped and still possesses large tracts of natural environment. The pace of occupation in these areas has grown exponentially with population. This not only has put long-term pressure on land resources, but has also forced the pace of road construction through the region. The importance of roads can hardly be exaggerated: they frequently represent the sole means of transport but also thread across many of the remaining fragments of undisturbed landscape. Whilst transporting people and goods, roads also act as corridors for pests, diseases and the panoply of environmental damage which accompanies occupation and settlement. The impact of the trans-Amazônia, Belém–Brasilia, São Paulo–Acre routes has been well described in the literature. Although the complete projection for the Brazilian trans-Amazonian perimeter road is unlikely to be fulfilled in the foreseeable future, it has already had a profound effect on the environment. Roads act as the precursors of development and provide the means for both the arguable benefits and deficiencies of colonization. Very little land development has so far proved to be sustainable and the momentum is largely maintained by political pressures.

The construction of the Pan American highway and its connecting road network through the Andes, or the building of the Atlantic coast roads through Mexico have produced analogous problems. Roads progressively shred the natural environment; they bring both civilization and destruction in their wake.

Political and strategic policies

Political and strategic issues are never far from all development programmes. Openly or covertly, they determine much of the opening up of frontier lands. A good example of development closely related to political ambition is that of the Calha Norte in the Brazilian Amazon.

There have been numerous government projects and development programmes in Amazonia going back to the late 1960s. These have been both economic and political in character. The economic motives have been said to follow the assumptions and directives of modernization theory, optimistic of a growing market economy and consequent improvement in social conditions. However, the Calha Norte had a different motivation and originated in military rather than economic sources in government. The idea was that the vast region to the north of the Solimões–Amazon rivers should be better integrated into national life by 'overcoming the great obstacles posed by the environment'. Cutting through the rhetoric, the project may be seen as a means of enhancing national security and reinforcing sometimes dubious claims to territory, as a means of consolidating frontier settlement, reducing smuggling, drug trafficking and existing or potential guerrilla activity.

The controversy centres around the protection of the considerable Indian communities and their environment on the one hand, and federal intervention in settlement, military installations and (unsympathetic) economic development on the other. The early development of the Project was not helped by its secretive nature. Over 80 per cent of the resources went to the military and there was the view, understandably, that this lack of balance and form of development was not compatible with the best interests of the inhabitants or future economic direction.

Conclusions

The environment currently receives considerable regional and global attention (namely UNCED 1992), but this is not translated into effective action on the ground, and lessons from the past are not all learnt. It is argued that the current development model is unsustainable.

Outside financial interests predominate in opening up frontier areas or consolidating existing settled regions. Satisfactory regulatory controls exist, but lack of funding and cross-institutional co-operation make such controls difficult to monitor and enforce.

Many of the countries are very large and developments affecting the environment are remote from the centres of power. Many of the economic pressures driving the pace of development lie outside the region. In addition, many regions are effectively small landowning or commercial fiefdoms with little interest in preserving the environment.

Despite this pessimistic view, the NGOs, national organizations and international environmental movements are growing in numbers and influence. With a still-favourable land-to-population balance in rural Latin America, there is time to evolve policies which can protect much of the present environment for future generations.

Further reading

FAO (1993) *Forest Resources Assessment 1990: Tropical Countries, Forest Paper 112*. FAO, Rome. (The most recent of the FAO overviews of forest resources dealing, for the first time, with the different types of forest reserve.)

FAO/UNEP (1983) *Guidelines for the Control of Soil Degredation*. FAO, Rome. (Dated but still useful outline of the methods used in controlling soil loss and land impoverishment.)

Furley P A and **Newey W W** (1983) *Geography of the Biosphere*. Butterworths, London. (Assesses the nature of the major global ecosystems, explaining their principal biological and environmental characteristics.)

Hall A (1989) *Developing Amazonia*. Manchester University Press, Manchester. (Examines the development of the Grande Carajas project in north-east Brazil within the broader context of the Amazonian environment.)

Prance G T and **Lovejoy T E** (1985) *Key Environments: Amazonia*. Pergamon Press, Oxford. (The most complete environmental and biological account of the Amazon region, which also includes aspects of land use such as agriculture and forestry.)

Whitmore T C and **Prance G T** (eds) (1987) *Biogeography and Quaternary History of Tropical America*. Clarendon Press, Oxford. (Comprehensive treatment of the theories for forest contraction to refugia during the periods of Quarternary ice advance, with a discussion of current understanding about the evolution of vegetation formations.)

World Resources Institute (1994) *World Resources 1994–95*. Oxford University Press, Oxford. (The best overall synthesis of resource data currently available, with chapters analysing many of the key environmental issues affecting Latin America.)

CHAPTER 5

Rural development: policies, programmes and actors

Anthony J. Bebbington

Introduction

The idea of rural Latin America conjures up a host of different images. Quechua and Aymara people growing native crops on the steep slopes of the Andes; the Amazonian forest occupied by indigenous hunters, gatherers and shifting cultivators, but being destroyed by large and small colonist farmers; semi-arid lands in the interior of Chile, the source of increasingly popular wines in Northern supermarkets; livestock in Central America caught up in the famous 'hamburger connection'; plantations and large rural estates. The list goes on. These are images of the rich and the poor, the indigenous and the modern, the sustainable and the destructive. The geography of rural Latin America is characterized by diversity and difference.

Not only difference, however. As the opening chapter suggests, there are common processes that cut across this diversity. One is integration into a world economy. Amazonian Indians selling forest products to the Body Shop and transnationals selling bananas from Central America both trade in a global market. That trade brings change to the rural areas where the commodities are produced. Latin American countries have also followed, to a greater or lesser extent, similar patterns in economic policy, the effects of which have reached down into rural areas. This continues through to today, with the continent-wide adoption of neo-liberal policies, the reduction of government support to rural development, the removal of subsidies and price controls, and the increased scope for private activity.

Linked to these processes of integration and globalization, one of rural Latin America's most significant shared experiences in the twentieth century has been the idea of 'development'. With the help – or at the behest – of Northern states and multilateral funding agencies, the governments and élites of Latin America have tried to promote the 'development' of their countrysides. This development is another reflection of globalization: the models of development pursued have usually been based on ideas from the North, and the resources channelled by Northern governments ultimately depend on the taxes paid by Northern citizens and companies.

Alongside these dominant, government-centred development programmes, other organizations have pursued their own ideas of rural development. NGOs, organizations of rural people, the churches and many others have all sponsored programmes or pursued objectives through their own activity and intervention in rural Latin America. These initiatives have introduced diversity into development debate and practice. But they too are linked to the North. Some of the money spent in Oxfam shops, or given at church finds its way to Latin American NGOs and church organizations. Without that money, many of these Latin American initiatives would not have been possible – or at least would not have taken the form that they took.

The very diversity of organizations aiming to bring change to rural Latin America suggests that there is no single 'rural development'. Rather, different actors have had, and continue to have, different conceptions of what rural society and economy should look like: sometimes the focus has been on modernization, sometimes on poverty alleviation, sometimes on growth, sometimes on environmental sustainability, sometimes on indigenous people, and so on. But in all cases, these programmes have occurred *in context*: that is to say, within a particular local context, a particular national and policy context, and a particular political and ideological context. Those contexts have inevitably influenced the ideas in those programmes, and the impacts of the programmes.

My objective in this chapter is to consider some of these approaches to rural development, and to investigate some of the links between the nature of programmes, their impacts in rural areas and the wider context within which they have been conceived and implemented. There is no single story to tell about the geography of rural change in Latin America: there are many stories. The one thing that links these stories together is that these different approaches to rural development, and the diverse contexts within which they are enacted, have continued to reproduce a rural Latin America that remains remarkably diverse, and yet at the same time is increasingly integrated into broader, common and indeed global patterns of change.

So on the one hand this chapter emphasizes this diversity, and is particularly concerned to show that ideas about, and experiences of, rural development have always been contested, and have differed from place to place and among different social groups. On the other hand, and despite this emphasis on diversity, the chapter also has an argument, and it is this. Much of what has passed as rural development in Latin America has been a reflection of integration and globalization. By and large, the achievements of these programmes have fallen well short of what was hoped for. These shortcomings have in turn prompted a search for alternatives. In some sense, these are specifically Latin American alternatives, based on local experiences and ideas developed within Latin America. Of course, these alternatives are today being pursued within a broader policy framework that is global in origin: the framework of liberalization, free trade and public sector cutbacks that we also know in the North. Nonetheless, under certain conditions these Latin American alternatives could have the potential to open up new opportunities for a more sustainable and democratic rural development. One of the challenges of the 1990s and the beginning of the next century is to go beyond the rhetoric of sustainable and democratic rural development, and to make it a reality through practice.

To tell these stories, the chapter opens with some history: that of change in Latin America before the Second World War. We will then look at the ways in which Latin American governments promoted rural change this century, and some of the impacts this has had on overall patterns of land use and landholding. This will lead to a review of some of the ways in which these government programmes have been questioned and resisted: particularly around questions of technology, equity, environment and ethnicity. We finish looking at the contemporary scenario for rural development under neo-liberalism and democracy, and consider some of the directions that rural development may take in the next few years.

Modernization and capitalism in rural Latin America

From the colonial legacy to import substitution

As outlined in Chapter 2, the colonial relationship left its mark on rural Latin America. As the economy of the rural estates, the so-called haciendas, steadily expanded through the seventeenth, eighteenth and nineteenth centuries (albeit with ups and downs), so they grew in size. In this way they exercised control over land and therefore over labour. As labour was scarce, estates created various means through which they could tie labour to the estate economy. In some cases, the estate had the right to a number of days' labour in return for granting access to pastures, water, fuelwood or other resources. In other cases, labour was tied by debts to the estate owner that the (usually indigenous) labourers were in effect obliged to incur. And often, families had to work on rural estates in return for access to land. The hacienda economy was thus linked to a small farm, or *minifundio*, economy which gave it labour.

Whilst the hacienda economy was largely oriented toward domestic markets, and was not particularly dynamic, a more dynamic, export-oriented and entrepreneurial rural economy also developed in the form of plantations producing export crops such as cocoa, coffee, sugar and bananas (for instance in parts of Peru, Ecuador, Venezuela, Brazil and Central America).

In painting this picture of a plantation/hacienda–*minifundio* political economy, it is important not to over-generalize. Historical research has also shown that there was much variation in how haciendas operated, and in the extent to which haciendas dominated rural areas. Similarly it has become clearer that in some areas, such as the central Andes of Peru, there was already a dynamic medium-sized farmer economy. Nonetheless, it is reasonable to say that the main heritage of the colonial period was a highly unequal landholding structure in which a white or *mestizo* (Chapter 6) minority owned most land, and Indians had access to very small amounts of land.

The export orientation of Latin American agriculture increased in the second half of the nineteenth century, in response to increased demand for foods from an industrializing Europe and North America. Some authors calculate that Chilean wheat exports increased 14-fold between 1845 and 1895, and Brazilian coffee earnings grew 13.5-fold in the period 1832–1870. Northern capital also invested in these ventures. Growing exports and investment led to further growth in the size of some export-oriented rural properties. However, competitiveness still came more from cheap labour rather than from investment in new technology: of this there was little.

The bubble to this growth burst when the Great Depression hit Europe and the USA in the late 1920s and early 1930s. Export markets collapsed. With this collapse, export-oriented development was no longer viable, and Latin America had to find a new development model.

The new model that emerged became known as import substitution industrialization, or ISI (Chapter 9). If Latin America was not going to be able to export agricultural and other primary products to finance its purchase of manufactured goods, it would have to develop its own manufacturing and industrial base. This was to be done through direct and indirect state influence over, or involvement in, investment in industry and infrastructure. This was coupled to trade policies to protect that youthful industry from competition.

The ISI model had many implications for rural areas. Agriculture was cast a subordinate role to industry – but this was still a critical role. Governments looked to agriculture for the production of cheap food for urban workers (in order to keep wage costs down). But policy makers also looked to the agricultural sector to generate profits which could be transferred to the industrial sector. Potentially these were two contradictory objectives.

State policies and rural modernization

These two pressures meant that it was imperative that the agricultural sector had to modernize and become more efficient. The state began to assume an increasingly active role in forcing this modernization. This was manifested in early investments in the provision of agricultural services, and then in land reform and integrated rural development.

These state policies undoubtedly influenced the face of Latin American agricultural society and the rural landscape. At the same time, however, the state maintained an overall policy framework that favoured industrialization and urban areas: in particular, it kept food prices low and concentrated investment in urban areas. So while Latin American governments carried out programmes to modernize agriculture, they also maintained a broader policy framework that would be an obstacle to this modernization.

Agricultural services

One of the principal ways in which Latin American governments promoted agricultural modernization was through investing in the provision of different support services to the rural sector. Credit was provided, irrigation infrastructure was installed, and institutions were built for the generation and dissemination of agricultural technology (so-called research and extension services).

Together these programmes made up the basic toolkit of the Green Revolution approach to agricultural development. As we will see later, there has been much debate about the social and environmental impacts of this approach – a debate that has nurtured alternative proposals for agricultural development in Latin America. What is clear, however, is that the approach helped transform Latin American agriculture.

The essential argument behind this Green Revolution model was that agriculture would only become more efficient and productive through the incorporation of new technologies – mainly modern crop varieties, fertilizers and pesticides. Linked to these new technologies would be the adoption of new practices, such as mechanization and irrigated farming. So, for instance, Mexico irrigated a million hectares of land in the 1950s. Various Latin American countries developed and strengthened national research systems to develop modern technologies. Much of this was done with support from Northern, particularly US, organizations who not surprisingly aimed to transfer Northern models of agricultural development to Latin America. The Rockefeller Foundation was one such important player, another was the US government. By the 1940s, the US Department of Agriculture was supporting agricultural research institutions in no less than ten different countries in Latin America. This US government support aimed to reproduce in Latin America the 'land grant' university model of the USA in which regional research institutions would deliver technological support to the farmers in the areas surrounding them through the work of agricultural extension services – although in Latin America the model was not necessarily implemented through universities and many of the research institutes were parts of government.

The response of the agricultural sector to these different initiatives was patchy and disappointing. Of course, there were pockets of modernization and change, but aggregate levels of productivity in many areas, and in many crops, particularly food crops, remained low in the 1950s and 1960s. The extent of technological innovation was limited.

Land reform

At the same time as aggregate increases in productivity were limited, social tension was increasing in the countryside. One of the critical origins of this tension was the pattern of highly unequal landholding. Of course, there had already been a long history of Indian and peasant uprisings in Latin America, as well as a long history of more mundane resistance to domination by white and *mestizo* élites, but these mobilizations became more co-ordinated and sustained in the twentieth century. Some came earlier than others, and led to significant political change. For instance, Mexico's revolution came in the 1910s, and Bolivia's in the early 1950s. These were 'revolutionary' movements that overturned government and enacted the wide-ranging land redistribution programmes that led to Mexico's *ejido* system and Bolivia's system of peasant unions, or *'sindicatos'* (though in both cases the peasantry sooner or later lost much control over the new state). By the late 1950s peasant movements with support from the political left, were pressing for similar programmes of land redistribution in other countries of Latin America.

By the 1950s, Latin American governments and élites were therefore faced with two concerns: to increase agricultural productivity and to diffuse growing rural conflict. These concerns opened the way to a round of land reforms that directly or indirectly changed the structure and culture of land ownership in much of the continent.

Many argued that the stagnant agricultural sector was a brake on Latin American development. They went on to argue that the source of this stagnation was the existing landholding structure of large estates. Data suggested that land productivity on large estates was far lower than that on *minifundios* (Tables 5.1 and 5.2). A series of studies by the Inter-American Committee on Agricultural Development argued that hacienda owners:

Table 5.1 Proportion of land cropped for different sized haciendas, Chimborazo Province, Ecuador, 1961

Size of hacienda (in hectares)	Percentage of area under crops
100–199	32.91
200–499	24.31
500–999	18.91
1,000–2,499	10.14
over 2,500	8.30

Source: Sylva (1986)

Table 5.2 Proportion of land cropped for different farm sizes, Province of Chimborazo, Ecuador (1954)

Farm size (hectares)	Total area (hectares)	Area cropped Hectares	Percentage of farm
0–4.9	52,300	49,900	95.4
5–49.9	48,600	31,900	65.6
over 50	214,700	37,500	17.5

Source: adapted from a report by Emil and Wava Haney (see their chapter in Thiesenhusen 1989)

121

1. Invested little capital in their farms.
2. Used their profits for conspicuous consumption rather than investment in new technology.
3. Held land as a status symbol, rather than for productive purposes.
4. Continued to use bonded labour – which, because it was free of charge, also discouraged any technical innovation.

These same proponents of land reform argued that small farms showed a far more efficient and productive use of land. This led to the claim that a redistribution of land into smaller units would increase aggregate productivity levels. It would also, they claimed, lead to income redistribution to small farmers who would then purchase more of the products of urban industry. Among the most optimistic, land reform was expected to achieve growth, equity and efficiency at the same time.

At the same time Latin American governments and the US government were also worried by the revolution that brought Fidel Castro and communism to power in Cuba. They feared that communism might be exported to other Latin American countries if sources of social tension, and in particular unequal land distribution, were not addressed. Under a programme called the Alliance for Progress, the Kennedy government made funds available to Latin American countries on condition that they removed institutional obstacles to democracy. Land reform was identified as a priority. That was 1961. The early 1960s then saw a flurry of land reforms until 1964 (Table 5.3). Once again, US influence was brought clearly to bear on the region.

Table 5.3 Selected land reforms in Latin America

Year legislation passed	Country
1960	Venezuela
1961	Colombia, Costa Rica, El Salvador
1962	Chile, Dominican Republic, Guatemala, Honduras, Panama
1963	Paraguay, Nicaragua
1964	Peru, Ecuador, Brazil
1969	Peru
1973	Ecuador

On the one hand, land reform legislation in the 1960s was a continent-wide phenomenon. It responded to general pressures related to capitalist development, and to hemispheric politics. It was clearly a phenomenon associated with integration and in some sense globalization. But on the other hand, different land reforms took various forms. Some reforms passed land to individual farmers, others to groups of farmers. In some countries reform was implemented very strenuously, in others it was not. In some countries significant areas of land were redistributed, in others redistribution was minimal (Table

5.4). Two variants of the land reform experience are discussed in Box 5.1.

Table 5.4 Land redistributed under selected land reforms in Latin America

Country	Percentage of forest and agricultural surface affected	Percentage of farming families benefited
Bolivia	83.4	74.5
Chile	10.2	9.2
Colombia	9.6	4.2
Ecuador	9.0	10.4
Peru	39.3	30.4

Source: adapted from Thiesenhusen (1989: 10–11)

Box 5.1:   Two paths of land reform: Peru and Ecuador

Land reform began in Peru as a response to peasant mobilizations and their invasions on to the land of large estates in the late 1950s and early 1960s. This mobilization was particularly strong in an area called La Convención, in the highlands. The first land reform legislation was passed in 1964. This legislation was not very far-reaching. It was only applied in the areas of greatest conflict, and in areas where pre-capitalist labour and landholding relationships continued. It transferred only 4 per cent of Peru's agricultural land, and a mere 2 per cent of the peasantry benefited. What land was transferred was given to individual families.

In 1969, however, a second, more profound, land reform was enacted by a radical military government. This reform was part of a larger programme which promoted co-operative and collective forms of production in Peru, and a model of development in which the state played a central role. Some 39 per cent of the land registered in the 1972 census had been redistributed and around 30 per cent of families had benefited. Over 95 per cent of the expropriated land was transferred to co-operatives, communities or groups. Very little was passed to individuals.

The reform increased equity in the countryside and brought an end to feudal forms of production. It also increased the role of the state in the countryside. However, the co-operatives it created did not function well, and agricultural growth lagged behind population increase. This was due to many reasons: management problems in the co-operatives; a cheap food policy which undermined incentives to agricultural production; and government control of the profits of the co-operatives which also undermined incentives to increased production.

The economic problems of the co-operatives led a new, right-wing government elected in 1980 to pass legislation to allow the subdivision of the co-operatives. The individual forms of land ownership that this promotes may well lead to increased social differentiation in the countryside, with the emergence of a capitalist family farm sector, and a growing landless rural labour force.

Like Peru, Ecuador has also had several stages of land reform, but the pattern taken by land reform neither promoted collective forms of production, nor distributed as much land. As in Peru, the first land reform in 1964 was weak and concentrated on abolishing feudal labour relations. A 1970 law for the coastal area abolished sharecropping relations. Very little land was redistributed in this period.

123

Box 5.1: Two paths of land reform: Peru and Ecuador
(continued)

Peasant organizations increased their pressure for land redistribution in the 1960s and this led to a further land reform law in 1973 that deliberately aimed to control this protest and promote economic modernization. In the words of the government's 1973–79 development plan:

The root of the problem is the concentration of land; the elimination of this concentration, and of precarious land tenure relations and other forms of exploitation, will permit an income redistribution which will automatically increase the purchasing power of the peasantry, creating a far wider market for popularly consumed industrial products.

The 1973 law was, however, far more effectively resisted by the organizations representing larger farmers than in the case of Peru. The government was also weaker and less radical than the Peruvian government and so less committed to the reform, and less able to carry it through. In 1979 the reform was essentially suspended. Less than 10 per cent of the land area was affected by the reform, and only 10 per cent of families benefited. Most land was passed to individuals. However, what did happen is that middle-sized farms modernized rapidly in the 1970s, using financial and technical assistance that they won from the state as a concession to their tolerance of some land reform legislation. The land reform thus helped trigger a modernization of significant sectors of Ecuador's rural economy.

By 1994 the medium and large farm sector was so in control of state policy that the Ecuadorian government attempted to pass legislation that would facilitate the private sale of land that has traditionally been held by indigenous communities.

In countries with Amazonian lands, reform programmes were also closely linked to programmes of agricultural colonization. These programmes gave incentives to small farmers and landless people to leave their home areas, and to settle (or 'colonize') so-called unused land in lowland forest, scrub and grasslands (though often that land was in fact the traditional territory of indigenous peoples). Indeed, to send people to the Amazon was one way of giving land to the rural poor without having to take much away from existing owners. It was also a way of diffusing land conflicts, and of trying to bring new land into production. From the 1960s onwards, the Amazon basins of countries such as Brazil, Peru, Ecuador and Bolivia all saw a number of large-scale colonization schemes. Subsequently, these schemes led to some of the types of damaging environmental and social impacts discussed in the preceding chapter.

Most of these land redistribution programmes enjoyed only a short life. In some cases, they were limited to the areas where conflict had been greatest – a clear reflection of their political intent to diffuse rural protest. In other cases such as Chile and Peru, where land reform policies were linked to changes of ruling regime and attempts to transform society, they and the political transformations to which they were linked encountered resistance from dominant interests and land-holding groups. In some cases this resistance was violent and at its most extreme, as in Chile, the resistance helped usher in a military coup that immediately overturned reform legislation, and subsequently returned some land to its initial owners. In the Peruvian case, the resistance was less violent but nonetheless played an important part in bringing the so-called 'Peruvian experiment' to a halt (see Box 5.1).

How should we understand this period of land reform then? Clearly the experiences were quite diverse, and therefore to make a generalized interpretation is risky. However, some things seem clear. Land reform was motivated by a dual concern to diffuse social protest and to modernize the agriculture sector. In many respects it was successful in both senses. Even in cases such as Chile, where land was returned to some of its initial owners, the old feudal oligarchy was dealt its final death blow by land reform. This oligarchy either lost land, or modernized its production process in order to avoid having its land taken away. At the same time, the transfer of land to reform beneficiaries brought more farmers into the market: they became potential producers of marketable crops, and potential purchasers of commodities. Meanwhile, on the other side of land reform legislation, many peasant movements emerged weaker: repressed in cases such as Chile or Guatemala, and left without an agenda in those cases where land had been redistributed.

Land reform was thus part of the transition to capitalism in Latin America. But this is too simple an interpretation. The forms taken by this transition differed across space, and so land reform contributed to sustaining the diversity of rural Latin America. Some areas emerged with a small farm economy, others with an agricultural economy based on medium-sized, modernized farms.

In many cases the transition remained incomplete. The new generation of small-scale, landholders were no longer tied to feudal estates, but nor were they yet modern, capitalist farmers. The next stage of government-led agricultural development was therefore an attempt to make them just this.

Agricultural and integrated rural development
Whatever the final verdict on land reform it was an important stage in the continuing integration of small producers into the market. Once producers had land, Latin American governments began to assist them with additional services to introduce modern technology to the small farm sector. Between 1962 and 1968, Latin American expenditure on agricultural technology research doubled; expenditure on agricultural extension to take that technology to farmers more than doubled. In the 1970s the World Bank identified integrated rural development as a priority programme.

Small producers had become an object of national policies. Governments looked to them to play an important role in increasing domestic food production. In pursuit of this goal, governments began to launch so-called integrated rural development programmes (IRDPs). In areas of colonization, rural development programmes aimed to provide the infrastructure and support services necessary to integrate colonist farms into the national market. At the same time, these programmes aimed to diffuse tensions in areas where there was continuing conflict over land. IRDPs aimed to increase yields on small peasant plots and thus reduce or avert remaining pressure for land reform.

The IRDPs were 'integrated' in that they attempted to address problems of rural underdevelopment on several fronts at the same time: they argued that prior programmes which had relied only on agricultural technology, or only on popular education and community organization, were bound to fail because they addressed just one, isolated, element of rural poverty. Thus IRDPs aimed to bring packages of modern technology, credit and infrastructure to the rural areas. Sometimes they also incorporated social components such as health care, education and organizational strengthening. After the more radical ideas of land reform, the IRDP approach to rural development reflected a more conservative means of addressing constraints on small farmer production without affecting tenure relations.

The general assessment of the impacts of IRDPs has been cautious. Evaluations have often stressed that they failed to address the causes of rural underdevelopment and poverty, and as a result have been unable to achieve very much – though of course the IRDPs never really intended to address many causes of poverty, such as unequal distribution of land. At the same time, they have been administratively complicated, because they have tried to co-ordinate so many components at the same time.

Agricultural modernization revisited: the Green Revolution debate in Latin America
Whether the state was delivering agricultural services, conducting land reforms or implementing IRDPs, the common thread running through these policies was the idea that agriculture had to be modernized: both peasant agriculture and larger-scale agriculture. The assumed basis of this modernization was technological change. Land reform tried to remove barriers to technology change, IRDPs and other research, extension and credit programmes tried to bring new technology and the means to adopt it to the farm gate. But was this technology appropriate for small farmers? Indeed, was it appropriate for the national economies of Latin America?

The debate on the impacts of this modern technological package has been heated, and it goes back to the very origins of the Green Revolution. One debate which occurred in Mexico in the early 1940s and which involved a Latin Americanist geographer is particularly interesting. It is an early example of debates that still rage between those who argue that Latin American agricultural development should be based on the introduction of modern technologies, and those who argue that development should be based on technologies that are indigenous to Latin America and that are ecologically sound.

The Rockefeller Foundation, a US foundation that still provides significant support to agricultural development, at that time wanted to support the development of agricultural research in Mexico. In the early 1940s, as it was considering how best to do this, the Foundation hired Carl Sauer to advise on the strategy and organization of this programme. Sauer was a cultural geographer at the heart of the so-called Berkeley school, and was particularly interested in the relationships between society and the environment, and in the ways in which local populations adapted to their environment. Sauer therefore argued that a Mexican agricultural development strategy should not be based on the introduction of technology based on Western science. Rather, he argued, agricultural development should build on the knowledge of Mexican campesinos (peasant farmers). Only this way would local cultures and local technological adaptations to the natural environment be maintained; only this way would technological innovations meet campesino needs and be sustainable (although Sauer did not use the language of 'sustainable development', a concept that was to be popularized far later).

The Rockefeller Foundation had a different analysis of the food question in Mexico, and saw campesino agriculture as technologically stagnant and backward. It therefore rejected Sauer's arguments, and concluded that Mexican agriculture required wholesale modernization. At the same time, it seems that Rockefeller felt that the main challenge was to increase food supply in whatever way. This implied that agricultural development and modernization programmes should not concentrate only on the small farm sector.

So Rockefeller continued with its programme of establishing modern research capacity in Mexico to develop modern crop technological innovations. This strategy was an important part of the Green Revolution in wheat (and to some extent maize) technology, that affected Mexico and many other countries. The Rockefeller Foundation was also providing financial and technical assistance in Colombia (1950), Chile (1955), Ecuador (1956), Peru (1956) and to Venezuela by 1960. At the same time it and other North American organizations supported the professional formation and training of many scientists in Latin America – studying in the USA they were therefore trained in an environment in which the necessity of agricultural modernization was largely taken for granted. Through this support, the Green Revolution approach – solving social problems with new technologies – was propagated, and institutionalized. It was also institutionalized in the form of International Agricultural Research Centres (created in Mexico in 1959, in Peru in 1970, and in Colombia in 1967), and in the national technology development programmes to which they still give support.

By the 1960s and 1970s, a critique of the Green Revolution in Latin America was gathering momentum. Several pieces of work began to suggest that larger farmers benefited most from the Green Revolution in Latin America. Because the research that went into developing these technologies was conducted under favourable conditions, they were more adapted to the farming systems of those wealthier farmers who owned high-quality, level land with irrigation. They performed far less well on the stony, sloping, unirrigated land that campesinos typically own. Furthermore they needed additional inputs of fertilizers and pesticides if they were to perform well: campesinos had less access to these inputs and to the credit needed to purchase them. In other cases, larger farmers used their political contacts to ensure that subsidized credit and inputs went to them. This bias to the large farmer meant that the Green Revolution package often fostered further social inequality in rural Latin America.

At the same time, the Green Revolution path has been questioned on the grounds that it did not and still does not make macroeconomic sense. It used capital intensively, though this was a scarce input (or 'factor of production'). At the same time it sometimes reduced the use of abundant factors of production such as labour. Labour was displaced by machinery. In this way, the landless lost out twice in this approach to agricultural development. Rural employment opportunities were reduced; and as they were landless, they had no opportunity to use any of the Green Revolution package in their own production.

More recently the Green Revolution package has been called into question on ecological grounds. Agrochemicals contribute to environmental degradation, polluting waters, and damaging the local microfauna that serves as a natural predator on crop pests. Similarly, chemical fertilizers do not return organic matter to the soil and so do not address underlying problems of soil degradation – indeed they may contribute to the further degradation of soil quality and structure.

Finally, agricultural modernization has also been questioned on cultural grounds. Following in Carl Sauer's footsteps, social scientists in Latin America have for some time expressed concern that modern technologies can displace native technologies and so displace an element of material culture in local societies. More recently, increasingly strong organizations of indigenous peoples have also begun to raise these concerns.

In many regards, these critiques merely echo what Sauer was warning 50 years ago. Of course, each is contested by those with a more sympathetic approach to the Green Revolution, who would argue that things have not been as bad as some critics argue. Indeed, it is also the case that in some instances the rural poor have taken advantage of modern techniques, using them in a scaled down form as part of a relatively sustainable rural livelihood strategy. Whatever the correct interpretation, once again this highlights the diversity of ideas regarding an appropriate agricultural development in Latin America. In the next section we consider some of the alternative approaches to rural development deriving from these critiques.

Lessons from state-led agricultural modernization
As can be seen from the previous sections, the period of state-led modernization has been subject to various forms of criticism. There are perhaps three of these criticisms that have had particular significance for subsequent thinking about rural development strategies:

1. One lesson of these experiences has been that government institutions are constrained in their ability to implement policies effectively – for political and bureaucratic reasons. They are subject to influence by political interests, are sometimes corrupt, have developed top-heavy 'development' bureaucracies and have often been inefficient in implementing programmes. With the cutbacks on government spending that have occurred in the 1980s and 1990s, this inefficiency may well have worsened.
2. A second, very influential, criticism has been that the modern technological package of new varieties, fertilizers, pesticides and machinery is inappropriate for small farmers and is environmentally and economically unsustainable.
3. A third criticism is that many government programmes have failed to have much impact because they have not addressed the causes of poverty. They have not addressed the dominant power relationships which tend to marginalize the poor from participating in development and political processes, and have not strengthened the capacity of the rural poor to address the causes of their poverty and political marginalization. The nearest that government policies have come to addressing some of these issues was in the land reform period, but even then government commitment was short-lived, and was blocked by more powerful interests.

These criticisms have had an impact on orthodox thinking and on the practice of development agencies. The more conservative – but still significant – response has been an increased agency interest in adapting programmes in order that the poor can participate in the modernization process. A more radical response has been to look for alternative (and not merely adapted) approaches to rural development. These are 'alternative' in the sense that they work outside the state, they work with other technologies, and they try to find responses to poverty that go beyond the provision of technology and inputs like credit. These alternatives are considered in the next two sections. The first section considers some of the alternative approaches that have been considered, and the second, the alternative actors.

Alternative agendas for small farm development

The period since the late 1970s has seen increasing discussion of different approaches to rural and agricultural development with a small farmer focus. Some of these alternatives are more radical than others. Some are more Latin American, others are related to international debates that have been brought into Latin America. The discussion of alternatives therefore reminds us of themes introduced at the beginning of the chapter: firstly, that there is much diversity in Latin American rural development – in this case, diversity of approach; and secondly, that the patterns of Latin American rural development reflect globalization – in this case, integration into global networks of ideas on rural development.

Reforming modernization: adapting institutions

There is no doubt that these critical reflections on rural development strategies have influenced the approaches of governments and orthodox development agencies in Latin America. This rethinking has not led to a wholesale rejection of the Green Revolution paradigm. Rather than reject the strategy of technological modernization, the emphasis among these organizations has been on adapting it. The argument here is that new technology is still essential for the development of campesino agriculture. It has had limited impact to date, not because the technology is fundamentally flawed, but for perhaps three main other reasons: firstly, campesinos have not had access to the necessary support institutions to help them adopt modern technology; secondly, the technology promoted needs further adaptation to campesino production systems; and thirdly, markets have been biased against campesino agriculture.

This diagnosis has led to programmes that have the following aims:

1. To enhance campesino participation in the generation of technology in order to ensure that technologies are more suited to campesino production systems.
2. To decentralize technology and rural development services, in order to ease campesino participation in their activities, and to bring institutions closer to rural people.
3. To increase the involvement of the non-governmental and private sectors in rural development, on the grounds that they are more able to deliver appropriate support to campesinos, are more efficient in doing so, and (in the case of NGOs) are more committed to working in the more remote and poorest rural areas which were often not reached by government services.

These concerns have been reflected in a flurry of programmes throughout Latin America promoting decentralization, participatory research and extension, and increased financial support for local government and NGOs in the provision of rural development services.

Questioning modernization: indigenous knowledge and agroecology

Criticism of the Green Revolution has also inspired more fundamental rethinking of approaches to agricultural and rural development. Increasing effort has been placed on identifying ecologically sustainable alternatives that do not depend on external inputs in the way that Green Revolution technologies do. In broad terms, there have been two interrelated approaches in this regard.

Many researchers and indigenous people's organizations argue that a sustainable agricultural development needs to be based on indigenous resource management and agricultural practices. The first step in such an approach to sustainable agricultural development is to understand indigenous practices, how they function and the logics that underlie them. The second step is then to find ways of improving them in order to intensify production without introducing external inputs.

In the Andes this approach has inspired efforts to recover terracing, indigenous irrigation systems and Andean crops such as *quinoa, tarwhi* and Andean potato varieties. In parts of Mexico, it has inspired programmes promoting raised fields (or *chinampas*) which allow intensive and organic agricultural production. In Amazonian areas, it has inspired efforts to intensify systems that do not imply forest clearance, and to link these systems with markets so that income per unit area is increased. Examples include the work with rubber tappers in the Brazilian Amazon, and the processing and export of nuts and other tropical forest products collected by indigenous groups occupying forest territory.

Not all these projects have been successful. Attempts to recover indigenous technologies from former periods have often encountered considerable resistance. This is because campesinos have usually had good reason to stop using these technologies. Attempts to reintroduce *chinampas* in Mexico, or terraces in the Andes, for instance, have gone awry when the group labour systems on which the traditional *chinampas* depended no longer exist, and when the profitability of production on raised beds and terraces did not justify the time spent in installing them. These experiences do not mean that such technologies do not have an important role to play in a more sustainable rural development. However, they do show us that 'indigenous' technologies may not always be appropriate for the current context of the campesino economy.

Another, and related, approach that has become increasingly influential in Latin America has been agroecology. Agroecology is an attempt to build agricultural systems grounded in ecological principles. Many of the origins of the agroecological movement, which now has international influence, in fact lie in the work of Latin American farmers, NGOs and researchers.

The agroecological alternatives being proposed for Latin American agriculture revolve around practices such as organic fertility maintenance, the management of ecological diversity, and the integration of crops, trees and animals in integrated production systems. Not surprisingly these approaches often look to indigenous production systems for inspiration. As this work has gathered momentum in Latin America, a continental consortium of organizations has been created to carry it forward – the Latin American Consortium for Agroecology and Development (CLADES). The pressure of these organizations, and of the environmental critique of the Green Revolution in general, has also led some of the international agricultural research centres to begin experimenting with agroecological approaches.

Much of this work is still experimental, but it suggests that an alternative technological agenda is slowly emerging in Latin America. This agenda argues that Latin America is now living in a post-structural adjustment world where the subsidies to Green Revolution technologies are being removed, and in a post-Rio world (i.e. following the United Nations Conference on Environment and Development which was held in Rio in 1992) in which environmental sustainability is on the agenda. In this new context, it is argued, agricultural development must be based on the intensification of indigenous, low external input techniques, rather than on the introduction of modern technologies derived from Western science. In this sense, the new agenda reflects a resistance to globalization, and an attempt to return to Latin American, locally derived alternatives for agricultural and rural development.

Challenging structure: the challenge of empowerment

But is technology alone sufficient to address problems of rural poverty and underdevelopment, *even if* that technology is indigenous and environmentally friendly? The history of Latin American rural areas suggests not. People are poor in Latin America because of skewed land distribution, unequal access to unfairly structured markets, weak organizations and dependence on external institutions. So, much more is needed than technology.

What is needed, say many, is power. If rural people had more power, they would be able to identify and address the causes of their poverty. They would carry forward their own development, rather than have models of development imposed on them by people and organizations who, however well-meaning they may be, are external.

So what is power, and where and what would it come from? This is a very difficult question – and potentially a theoretical one. However, for a long time there has been a current in Latin America that has stressed the need for so-called 'conscientization' and 'organization' as crucial elements of rural development. These ideas have influenced the rural development approaches of many Latin American non-governmental and church organizations. Indeed, like agroecology, conscientization is also a concept with strong Latin American roots of its own.

In simple terms, conscientization means people becoming aware of their reality and understanding what it is that makes them poor. An early proponent of this approach to 'empowerment' was the Brazilian educationalist Paulo Freire. It is an approach that people working in popular education (i.e. non-formal education with poor people) and liberation theology have long tried to promote among Latin America's rural (and urban) poor.

But once you have identified the causes of your poverty what do you do about them? This is where organization becomes important. Only through being organized, the argument goes, can rural people have sufficient power to address the causes of inequality, to negotiate with the state, to press for change and to carry forward their own development alternatives. The experience of the 1950s and 1960s, when

Latin American peasant movements had a real influence on their governments, suggests that there is truth in this. Consequently, many rural development organizations have tried to do such popular education work and link it to the creation and strengthening of organizations of rural people. Most of this work has been by NGOs and rural people's own organizations (see below).

Such empowerment work can be done in conjunction with efforts to recover indigenous forms of technology. To resist modernization in the ways outlined in the previous section can thus be a contribution to empowerment, and at the same time an outcome of empowerment. This is particularly so in the case of indigenous people's organizations for whom empowerment necessarily means enhanced pride in, and identification with, 'traditional' practices, which almost inevitably will mean a questioning of some aspects of modernization, and an endorsement of local alternatives. Furthermore, within the empowerment approach, to organize as a means of enhancing ethnic identity, and of restoring indigenous technical practices, may also be linked to the idea that organization is a means of questioning prevailing relationships of power. In Ecuador, for instance, the Centre of Peasant Education and Training for Azuay (CECCA) developed a project in this vein. In their explanation of the project, they say that the organization had as its fundamental goal:

> to achieve the autonomy of popular organization in the face of the state's efforts to control it.... [This autonomy would be promoted by recovering and promoting] a knowledge, a science, a culture, and a technology which has allowed it [the popular organization] to produce, to reproduce itself and to create a style of life in accord with its own cosmology. [This approach would permit the] search for alternative modes to the model of capitalist 'development', and this would in practice presuppose the genesis and consolidation of popular power.

This is a very different approach to rural development from the ones we have discussed so far. Rather than begin with the idea that rural people need *things* – technology, credit, seed, etc. – it argues that what is first needed is *power*.

However, the experience with these approaches has also been disappointing. What emerges is that what is in fact needed is empowerment *and* technology *and* organization *and* fairer access to markets *and* more accessible and more relevant services. This is a tall order. Are the so-called alternative actors of rural development capable of all this?

New actors for alternative agendas: from rural development to rural democratization?

These new agendas have primarily been developed by 'alternative' institutions outside the government sector. In particular, they have been developed through the work of non-governmental and church organizations, and the so-called popular, grassroots organizations of rural people. Who are these organizations, and how should we understand their emergence (Fig. 5.1)?

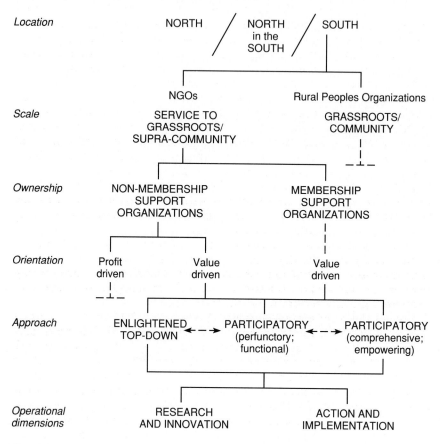

Fig. 5.1 Types of NGO and people's organizations

The Church and Non-governmental Development Organizations

At the turn of the century, large parts of rural Latin America were still controlled by the owners of rural haciendas and plantations. Together with the Catholic Church and the local state these institutions held sway over patterns of change (or stasis) in rural areas. Uprisings of peasants and indigenous people occurred from time to time, but power was in the hands of the estates and the Church (which was itself often a large land-owner). As we saw, the dominance of these institutions was progressively undermined as the twentieth century moved onwards.

As a result of land reform and rural development policies,

government began to replace the estate as the dominant institution in rural administration. At the same time other agencies moved in to fill the void left by the decline of the traditional estate. As they moved in, many of these organizations began to pursue programmes of rural development. One important such institution was the 'new' Catholic Church, which during the 1950s and 1960s in Latin America made a progressive commitment to the poor, and began to see social development programmes as a part of the theology of liberation. In other cases specialized non-governmental development organizations emerged, linked to the Catholic Church. In yet other cases, liberal and radical professionals committed to rural development created NGOs. In some cases these were people who saw land reforms stall, and became disillusioned with the possibility of achieving radical social change through the action of the state. In other cases, NGOs were formed when governments moved from liberal agendas to authoritarian and repressive ones, as happened at different times in Argentina, Bolivia, Brazil, Chile and much of Central America. These authoritarian regimes kicked liberals out of government, clamped down on left-wing political parties and repressed popular organizations. Creating an NGO was one of the few ways of trying to keep a liberal politics alive, and of giving some support to popular sectors.

These organizations received funds from different non-government, Church and governmental organizations in the North who sympathized with their alternative visions of rural development. Particularly during periods of repressive politics in Latin America, donor agencies supported the activities of these alternative actors.

Just as these organizations were diverse (Fig. 5.1), so they worked with a diversity of models. Some of those linked to the churches saw their work as social work and charity in rural areas; other, more radical, NGOs had a different vision of their work, which they saw as an effort to establish links with grassroots organizations, in order to strengthen the capacity of rural people to be politically active and to manage their own development. In this sense these NGOs saw themselves as a vehicle for rural democratization as much as for rural development. Not only were they delivering services, but they were also creating organizations. To use a term that has become fashionable in recent development jargon, they were strengthening rural civil society: they were increasing the capacity of rural people to protect their rights, to negotiate with the state, and to conceive and implement their own development activities. They were working within the 'empowerment' approach we outlined above.

In the short and mid-term, the theory was that the NGOs would be responsible for these actions themselves. In the medium and long-term these goals would be achieved by the rural people's organizations that the NGOs would help create. In this sense, the NGO was always supposed to be a transitory phenomenon. The theory was that it would be replaced by rural people's organizations. These popular organizations might choose to ask NGOs for advice, but the initiative would come from the people and not from the development NGO.

Has this happened? This is an important question, for it is only if this has happened that we can reasonably talk of rural democratization having occurred and of rural civil society being stronger as a result of the NGO's work. If it has not occurred, then the only thing that can be said to have occurred is that there are now more organizations involved in implementing development programmes. Rather than fostering rural democratization, this can merely lead to yet worse co-ordination of development activities in rural areas.

Rural people's organizations

Are rural people's organizations any stronger now than they were 30 years ago as a result of this support from NGOs and churches? Are they any more able to carry forward their own development, and negotiate with government?

The question is difficult to answer with any simple generalization. Thirty years ago was the heyday of peasant movements. Peasants were actively pressing for land reform, and the unrest in rural areas coming from peasant organizations was a critical factor in inducing land reform. But following the successes of land reform, many peasant movements became weaker. Their members had land, and the *raison d'être* for dedicating time and resources to the organizations seemed less clear to many members.

In the 1990s it is far from certain that rural people's organizations are as strong as these peasant movements were in the 1960s. At one level, many people argue that the macroeconomic policies of the 1980s, which aggravated the poverty of the very poor, had the effect of weakening popular organizations. This source of weakening was complemented of course by simple straightforward repression in countries such as Chile and Brazil. In some areas this repression still occurs.

It can also be argued that because NGOs have assumed the role of representing rural concerns to governments and donors, they have served to keep popular organizations on the margins of rural development. It does seem to be the case that the number of effective sustainable rural people's organizations that have emerged out of NGO activities are relatively few. It is also the case that the early years of the 1990s have seen increased criticisms of NGOs for not living up to their rhetoric and not being accountable to rural people's organizations. Far from empowering, Latin American NGOs are increasingly criticized for being paternalistic and unwilling to let go of power to grassroots groups.

In retrospect these criticisms may one day appear to have been excessive. The impacts of NGOs and other actors on the strengthening of rural organizations are not always direct and certainly not instantaneous. Similarly the pessimistic argument that rural people's organizations have become weaker may be an overstatement. And we know from our television screens and newspapers that there are strong organizations. Whatever else it showed, the armed uprising in Chiapas in southern Mexico in 1994 demonstrated that rural people still have

Plate 5.1 *Zapata Vive! Carajo!* Demonstration in Mexico City 1994. The no
longer revolutionary government of former President Salinas
changed Article 27 of the Mexican Constitution and allowed land
given to communities (*ejidos*) to be sold freely. The revolutionary
leader Emiliano Zapata doubtless turned in his grave but inspired
the Zapatista uprising in Chiapas, southern Mexico. (Photo: Tony
Morrison, South American Pictures)

significant power and motivation to organize and protect their
livelihoods. In Brazil, the rubbertappers hit our television screens when
their leader Chico Mendes was assassinated – and elsewhere in Brazil
rural unions have not all been repressed or swept aside.

One country in which rural people's organizations have certainly
become stronger has been Ecuador (see Box 5.2). While donors called
the 1980s the 'lost decade' in Latin America, the leader of the national
indigenous people's (or Indian) movement in Ecuador called it a decade
that had been won. The 1980s (and 1970s) in Ecuador saw the
emergence of Indian organizations at all levels: federations of
communities existing at county and provincial levels; a federation
covering the Amazon and another covering the Andes; and a national
level organization. As well as engaging in political action, these
organizations have their own rural development programmes. This is
one case where the rhetorical objective of strengthening rural people's
organizations so that they defend rural people's rights, negotiate with
the state and implement their own development programmes is nearing
reality (see Box 5.2).

Box 5.2   Indigenous people's organizations in Ecuador: organizing for rural development and rural democracy

Although indigenous people in Ecuador have a long history of organizing for specific uprisings, their organizations have become stronger and more formal in the twentieth century. Furthermore, over the last 20 years, the role of these organizations has expanded. Whereas historically they have been vehicles of protest and negotiation, more recently they have begun to implement their own development activities.

Earlier in the century, these organizations were mainly concerned with access to land, and were concentrated in the highlands. In 1944 the Ecuadorian Federation of Indians (FEI) was formed, followed in 1964 by the National Federation of Campesino Organizations (FENOC). These national organizations were primarily concerned with demands for land redistribution and land reform. Once land reform legislation was passed in the 1960s and 1970s, these organizations lost some of their momentum.

However, the period during and following land reform saw a different organizational process begin. NGOs, church-based organizations and even government integrated rural development programmes began to work with indigenous people, with the goal of creating new forms of local organization. In particular, they promoted the creation of regional federations of indigenous communities. Some of these federations have continued in their concern for land rights. In Amazonia, for instance, these federations have pushed hard to gain land titles for their traditional territories in order to protect them against the incursion of colonists and larger-scale agricultural enterprises.

Although the question of land redistribution is far from over in Ecuador, these organizations have also initiated other activities. They have negotiated with the state over rural educational policy. They have also become active in rural development. Sometimes this has been in partnership with other institutions; at other times, they have negotiated their own funding from Northern development agencies. With these funds they have designed and implemented their own development programmes.

These programmes have taken many forms: some with modern technologies, others with indigenous and agro-ecological approaches. The success of these programmes has been variable. Most programmes and organizations are still dependent on external funding, and the impact of the programmes on rural livelihoods has often been marginal. However, within this diversity the overall objectives of these programmes are relatively similar. They are an attempt to develop livelihood opportunities for indigenous people that will allow them to remain in rural areas. In this respect the programmes also aim to protect and revitalize indigenous culture – for this culture will only survive as long as indigenous communities remain strong, and are not undermined by migration.

The organizations and their programmes also reflect an attempt on the part of indigenous people to exert more control over rural development, and over rural administration in general. In this respect they have been more successful. Indigenous federations have become an important actor in rural and regional politics and are able to exercise significant influence over the rural state.

Coupled with the increased strength of regional federations, the national indigenous organization (CONAIE) has also become stronger. It has negotiated with government in several areas related to land rights, indigenous people's rights, education, agricultural policy and rural development in general. On two occasions in the 1990s it has called successful national uprisings of indigenous peoples. These uprisings lasted for several days and led to the closure of principal roads, obstructing national marketing of goods.

Box 5.2: **Indigenous people's organizations in Ecuador: organizing for rural development and rural democracy** (continued)

Today indigenous people's organizations in Ecuador are a significant actor in rural development who themselves are implementing programmes and elaborating proposals for alternative policies of rural change.

Again the lesson of this is diversity: the strength of rural people's organizations varies among countries: in some such as Ecuador they have become stronger; in others such as Chile and Peru they are now far weaker. Likewise some NGOs and churches have helped strengthen them, others have not. The extent to which we can speak of stronger rural democracy accompanying rural development thus also varies among countries.

What can be generalized, though, is that (a) these organizations have much potential as actors in rural development, (b) that rural development and progressive social change in rural areas will only be sustainable if these people's organizations become stronger, and (c) that these organizations can easily be weakened by repression and by macroeconomic policies that undermine their members' livelihoods.

Whither rural Latin America?

What does the future hold for rural development in Latin America? Prediction is always difficult, but there are dominant trends that can be identified in the region. On the one hand, they reflect the globalization of policy ideas in Latin America, and the increasing economic integration and modernization of the region. On the other hand, they will reflect the social, ethnic and environmental diversity of the region, and the continued efforts of local populations and organizations to create their own strategies and alternatives.

New policy agendas

One powerful measure of globalization in Latin America has been the forceful rise of neo-liberal economic policy since the 1980s (and since the mid-1970s in Chile). This policy agenda is characterized by the steady removal of subsidies, tariffs and other barriers that affect trade, by the reduction of state intervention in the market, and by cutbacks in state expenditure. In general, these are policies that aim to increase the scope for private sector activity.

This policy agenda has had a range of implications for rural and community development and small farmers. Two broad impacts are relevant here: those that influence the economic environment for agriculture, and those that influence the institutional and support environment for peasant farmers.

Among the main changes in the economic environment are the following:

1. Export possibilities increase because of currency devaluations, but competition from imports may also increase with the removal of tariffs and quotas.
2. Subsidies are also removed on agrochemicals making them more expensive and more difficult for poorer farmers to buy.
3. Subsidies on credit are reduced or removed – in some cases, such as Bolivia, the state agricultural banks have been closed.
4. Competition from, and access to, neighbouring countries' markets has increased as a result of the increasing integration of countries into regional trading blocks.

These changes have differing implications for agriculture. For instance, there is some evidence to suggest that areas producing export crops do benefit – conversely, areas producing crops with less export potential are not so likely to benefit. In Chile, for instance, local and regional economies linked to the forestry sector grew as a result of the promotion of an export-led development model. There has been much more debate, however, on whether the small farm sector would benefit from these changes. As this sector typically produces food crops rather than export crops, it is less likely to benefit from export possibilities, but will suffer from increased input prices and the loss of subsidized credit. On the other hand, another potential effect of this new policy environment may be to make the low input and agroecological technologies of indigenous agriculture more competitive with modern technologies, whose profitability may decline with the increased costs of agrochemical inputs.

It is probably too early to tell what the net effects of these changes will be on the welfare of small farmers. What is clear is that the progressive integration of small farmers into national, regional and global markets is going to continue in the foreseeable future.

Linked to these economic policy changes are a series of institutional changes. At the centre of these is the progressive reduction of the role of government institutions. The rationale for this reduction has been that:

1. Previous government activities in rural development have been inefficient and ineffective, and have stifled private sector initiatives.
2. With the policy commitment to reduce public sector expenditure, government institutions have to be reduced in size.
3. The private sector is more efficient in delivering rural services, and in allocating rural resources

These ideas have inspired efforts to shift many rural development responsibilities from government to the private and non-governmental sector. For instance, in Chile agricultural extension is now provided by NGOs, farmer co-operatives and private companies on contract to government. In Bolivia and Peru, in some cases the state simply does not provide these services and the government is effectively leaving these tasks to be implemented and paid for by NGOs. Throughout Latin America, special funds are being created to finance NGOs, community groups and private enterprises to implement rural development projects. The explicit justification for this is that these organizations are more effective, more innovative, closer to the rural poor and therefore better able to address rural poverty than is government.

This same reasoning has also inspired policies that aim to convert institutions that resist certain forms of market activity into ones that favour market activity. For instance, in some cases water legislation has been changed to create a market in water rights which were previously allocated by government or other non-market means. In Mexico, the *ejido* landholding system which prevented the legal sale of collectively held lands has been changed to allow market transfer of land; similar legislation has been considered in Ecuador to facilitate market transactions in the land of indigenous communities.

These different changes have – again – been greater in some countries than others. But the trend is clear. Increasingly government service provision is being down-sized and replaced by non-governmental and private provision; and legislation is being passed to diminish the strength of social and administrative procedures that allocate rural resources (land, water, forests, etc.), and to increase the extent to which these resources can be distributed through market transactions.

The end of the peasantry?

What do these new policies imply for the future of the peasantry in Latin America? For a long period there has been a steady shift of population out of rural areas and to urban areas (see Chapter 7 on migration, and Tables 5.5 and 5.6). This trend is particularly advanced in countries such as Argentina and Chile. These are now urban countries, with only a small campesino sector remaining. Their rural populations are increasingly of rural labourers rather than of rural producers. This trend is far less advanced in Central America and in the Andean countries of Peru, Bolivia and Ecuador which still have large rural populations who depend on agriculture for a significant share of their livelihoods. A large part of the rural population that is left in these countries is the campesino sector.

Table 5.5 The changing rural economy: percentage of labour force in agriculture

Country	1970	1975	1980	1985–88	1990–92
Bolivia	52	49	46	46.5	47
Chile	23	20	16	18.3	19
Colombia	39	37	34	–	10
Ecuador	51	45	39	38.5	33
Peru	47	43	40	35.1	35

Source: UNDP (1991, 1994)

Table 5.6 Changes in the rural population of Latin America

	Rural population (as % of total) 1992	Percentage of labour force in agriculture	
		1965	1990–92
Argentina	13	18	13
Belize	49	–	–
Bolivia	48	54	47
Brazil	23	49	25
Chile	15	27	19
Colombia	29	45	10
Costa Rica	52	47	25
Ecuador	42	55	33
El Salvador	55	58	11
Guatemala	60	64	50
Guyana	66	–	27
Honduras	55	68	38
Mexico	26	49	23
Nicaragua	39	56	46
Panama	46	46	27
Paraguay	51	54	48
Peru	29	49	35
Uruguay	11	20	5
Venezuela	9	30	13

Source: UNDP (1994)

Will this campesino sector survive? In an increasingly competitive agricultural economy, some, maybe many, campesinos will not survive. Furthermore, legislation facilitating the sale of land, such as that mentioned earlier in Mexico, increases the possibililty that campesinos will sell land to more commercial interests, and steadily the proportion of land under campesino control may decline. Certainly there is little legislation or political will left for land reform programmes in order to transfer land to campesinos. In many countries, land reform programmes have been frozen or reversed. The best that campesinos can hope for now is to be given legal title to land that has long been theirs anyway.

On the other hand, where will the campesino sector go? The assumption is that it will be absorbed into expanding urban labour markets. However, some question whether these markets will be able to absorb this number of new workers, and in some countries where the urban and manufacturing sector has not enjoyed economic growth on the same proportions as, say, in Argentina or Chile, then it is unlikely to be able to absorb these people.

Some governments and donors, such as the Inter-American Development Bank, are concerned that this unabsorbed mass of displaced rural people will be a potential source of tension, instability and unrest. Equally there is concern in some quarters that people who remain in rural areas, but remain marginalized from the dominant economic system, will also be increasingly vocal in their demands for

142

more secure livelihoods. The rural uprisings in Mexico's Chiapas and in Ecuador, and the rural violence of the guerrilla movement Shining Path (*Sendero Luminoso*) in Peru, have not been lost on policy makers.

These political concerns, coupled with the continued importance of the small farm sector in food supply, will mean continuing interest in rural development programmes for the foreseeable future, even given the long-term decline of rural population. However, these programmes will not look the same as the rural development programmes of the past. They will focus on improving the integration of rural people into the market, as farmers or as labourers; and they will be programmes implemented increasingly by the private and the non-governmental sectors.

A closing note: development democratized or privatized?

In this policy context, many NGOs and others who have developed alternative, Latin American visions of rural development have had to face the following dilemma: the arguments for privatization are consistent with what many of these so-called 'populist' alternative visions have been arguing for some time. The alternatives criticized government for being inefficient, bureaucratic and corrupt in implementing rural development; they also argued that the resources of many government rural development institutions did not reach the poor. They argued that government was too centralized.

The privatizers in Latin America would share all these analyses of the role of government in Latin American rural development. The principal difference is that the NGOs wanted a reformed and new state that would still play an important role in diverting resources to the poor, and in stimulating rural development. The privatizers, however, argue that the state should be reduced in size, and that where possible NGOs, grassroots organizations and the private sector should replace it in the provision of rural development services.

This presents NGOs and rural people's organizations with another dilemma. They long argued that they were better at rural development than was government. They are now being given the chance to show this, and are being offered contracts to implement programmes. On the one hand this will allow them the chance to have a greater impact, and have access to more resources. It might also allow them to feed some of the alternative, Latin American approaches to rural development into national programmes. On the other hand, if they accept these new relationships with government, they become party to the neo-liberal policies and public sector cutbacks with which they disagree, and they also find themselves more constrained by government procedures which can limit the extent to which they can experiment with alternatives. Some accept the new relationships; others are more cautious.

Different NGOs and popular organizations are responding in different ways, but it seems clear that the model of rural development for the rest of this century at least will be one of more partnership between NGOs, government, popular organizations and the private sector. These partnerships will imply an increased role and responsibility for these new, alternative actors. With that increased role may come the possibility of a greater chance to influence how rural development policies and programmes unfold. To this extent the privatization of rural development in Latin America does offer chances for the democratization of rural programmes at the same time.

But interpretations of this new agenda will continue to vary. Just as some argued that land reform programmes were progressive, so some will argue that the new agenda is a step forward towards a more participatory and democratic rural scene; and just as some argued that land reform was merely a way of easing a transition to capitalism, so some will argue that in the new agenda, NGOs will merely be subsidizing privatization, and so assisting in another capitalist transition in Latin American economies – a transition to a free market capitalism. In the future, as in the past, what constitutes an appropriate rural development strategy in Latin America will, like beauty, be in the eye of the beholder.

Further reading

Altieri M (1987) *Agroecology: The Science of Alternative Agriculture*. Westview Press, Boulder. (An overview, with many Latin American examples, of the agroecological critique of the Green Revolution, and the alternatives it proposes.)

Bebbington A and **Thiele G** (1993) *NGOs and the State in Latin America: Rethinking Roles in Sustainable Agricultural Development*. Routledge, London. (A review of the social origins of rural development NGOs, and their current work; locates these NGOs within wider patterns of social and technical change in rural Latin America.)

Bebbington A, Carrasco H, Peralvo L, Ramón G, Torres V H and **Trujillo J** (1993) 'Fragile lands, fragile organizations. Indian organizations and the politics of sustainability in Ecuador', *Transactions of the Institute of British Geographers* **18** (2), pp. 179–96. (A study of the rural development initiatives of indigenous people's organizations in Ecuador.)

Browder J (ed.) (1989) *Fragile Lands of Latin America: Strategies for Sustainable Development*. Westview Press, Boulder. (Case studies of different technological alternatives, many involving indigenous peoples and their technical knowledge.)

de Janvry A (1981) *Land Reform and the Agrarian Question in Latin America*. Johns Hopkins University Press, Baltimore. (This is a classic analysis of land reform programmes written from a neo-Marxist perspective.)

Fox J (1990) (ed.) *The Challenge of Rural Democratization: Perspectives from Latin America and the Philippines*. Frank Cass, London. (A collection of studies of rural social movements in several Latin American countries.)

Freire P (1970) *Pedagogy of the Oppressed*. Seabury Press, New York. (A classic book written from Latin America on concepts of empowerment and conscientization.)

Griffin K (1975) *The Political Economy of Agrarian Change: An Essay on the Green Revolution*. Macmillan, London. (One of the early studies of the biased effects of Green Revolution technology.)

Grindle M (1986) *State and Countryside. Development Policy and Agrarian Politics in Latin America*. Johns Hopkins University Press, Baltimore. (A thorough study of how land reform and IRDP policies were formulated and implemented.)

Hecht S and **Cockburn A** (1990) *The Fate of the Forest*. Verso, London. (A study of the history and political economy of rural change and rural development in the Brazilian Amazon.)

Jennings B (1988) *Foundations of International Agricultural Research: Science and Politics in Mexican Agriculture*. Westview Press, Boulder. (A polemical, but revealing, study of the origins and impacts of Green Revolution research in Mexico.)

Jordan F (ed.) (1989) *La Economía Campesina: Crisis, Reactivación y Desarrollo*. Instituto Interamericano para la Cooperación Agropecuaria, San Jose. (A collection of three very good overview essays looking at the peasant economy and approaches to rural development in Latin America in the post-1945 period.)

Long N and **Roberts B** (eds) (1978) *Peasant Cooperation and Capitalist Expansion in Central Peru*. University of Texas Press, Austin. (A study of peasant communities and livelihood strategies in highland Peru.)

Thiesenhusen W (ed.) (1989) *Searching for Agrarian Reform in Latin America*. Unwin Hyman, London. (A collection of case studies of different agrarian reform programmes.)

CHAPTER 6

Race, gender and generation: cultural geographies

Sarah A. Radcliffe

Generation, gender and race are fundamental divisions in the societies of Latin America, providing a basis for the creation of social identities and for the perpetuation of differences between individuals and groups. These three human dimensions mutually influence the uses of resources and spaces in the continent, and also account for varying benefits from the development process.

Race and place in Latin America

Race is one of the neglected areas of the geography of Latin America. Nevertheless, the ways in which different population groups make use of spaces and resources in the continent have long been influenced by the social meanings which are attached to racial differences. The social and political meanings of terms such as *'negro'* (black), *'mestizo'* (mixed race), *'gente blanca'* (white, or European) and *'indígena'* (indigenous, native Americans) have varied little from place to place throughout the continent. The term *racial formation* has been developed by Howard Winant in order to analyse the forms of interaction between racialized groups in Latin America. Looking at Brazil, he discusses how the historically and geographically variable racial relationships are a consequence of negotiation and debate between groups. A lack of clearly defined racial categories contributes to this situation, whereby the meaning and 'content' of racializing terms which ascribe people to a particular category can be changed with social reorganization. One feature of racial formations in Latin America is a racial labelling dependent more upon social context and personal attempts to 'pass' into another category, than upon physiological characteristics. As a result of the long histories of valuing 'whiteness' highly, there are various ideologies of 'whitening', which place higher value on lighter skin. During the nineteenth century, when European theories of eugenics and scientific racism were influential, several Latin American countries adopted policies to encourage the immigration of white

146

populations. In this century too, notions of *mestizaje*, or racial mixing (believed to unlock the potential for countries' development), have often contained undercurrents of seeking to whiten societies, but they have not resulted in laws on racial difference.

With this lack of strict 'racial' difference and the exaggerated valuation of whiteness, the social and spatial implications of racial difference are played out in very varied ways. It is the meanings of racial differences, and the institutionalization of practices concerning race, which must be understood in order to appreciate the dynamics of racial relations at different spatial scales in Latin America. We can examine some of these relations by looking at a variety of different geographical scales:

1. *At a regional level.* Certain regions of Latin America are arguably characterized by internal colonialism (see Box 6.1), whereby unequal socioeconomic relations are also overlain by relationships based on a hierarchy of racialized groupings. The classic example of this is Mexico, where the southern states have majority Amerindian populations which are impoverished and relatively powerless. Nevertheless, the idea of internal colonialism has been challenged by writers who argue that no specific administrative structure which carries through policies of underdevelopment can be found in these regions. Certainly too, the uprising of indigenous peasants in Chiapas, southern Mexico, in 1994 suggests that the conditions of internal colonialism are not passively accepted. Nor indeed are social actors in the 'internal colonies' solely concerned with issues at the local regional level: the peasant rebellion of 1994 was motivated in part by the Mexican government signing of the NAFTA (North American Free Trade Agreement) accord with the United States and Canada.

Box 6.1   Internal colonialism

From their work on the ethnically varied region of Cuzco, southern Peru, van den Berghe and Primov suggest that internal colonialism characterizes national regions which are in a dependent relation with the state. This internal colonialism has a series of racial elements:

1. The dominant and subordinate groups are ethnically different from each other.
2. The subordinate ethnic group is spatially segregated from other groups.
3. The subordinate group is subject to special and separate administration.
4. The subordinate group has a separate set of laws which relate to such specific concerns as communal land tenure (as occur in, for example, Mexico and Peru).

Source: **van den Berghe** and **Primov** (1977).

2. *In the city*. Residential segregation in many Latin American cities results in racial segregation, owing to the presence of racialized groups in distinct labour markets and with varying amounts of social, economic and political power. During the colonial period in Lima, the indigenous population was found only in one particular area, known as El Cercado. One study in contemporary Brazil examined degrees of residential segregation in 35 of the country's largest metropolitan areas, and found that whites are generally more segregated from blacks than from mulattos (mixed). The presence of significant residential segregation between blacks and mulattos, and a greatly increased rate of segregation with rising incomes, imply that socioeconomic status is not the sole factor. However, in low-income settlements there was a high degree of multiracial residence.

3. *In the home*. In the early nineteenth century, the separation of different racialized groups was based on notions of superiority and difference. It was, for example, unacceptable for descendants of slaves to sit with people who were 'noble, legitimate, white and free', according to a writer in 1806. Well into this century in Bahia, Brazil, low-income black groups could be received into élite homes but not sit in the living room or dining room; if a meal was offered, the guests would eat in the kitchen or at different times.

4. *On the body*. Manners of presentation and clothing codes have been used at various times and places in Latin America to mark racialized difference. Of course, similar markers indicated gender and class as well. During the colonial period, black women and men were not allowed to wear silk in Lima. Through much colonial and republican history, indigenous peasants were not permitted by the land-owners to wear urban-style dress.

Consequences of conquest

On the arrival of Spaniards and their armies in the Americas, the type of contact with the Amerindian populations depended on their political organization and sizes. On the eve of conquest, it is estimated that between 14.4 and 52.9 million people inhabited what is now Latin America, concentrated particularly in the Andean region (6.1–11.5 million) and the Mexican heartland (4.5–21.4 million). Where Amerindian populations were organized into states and were densely settled, as in Mexico and the Andes, the conquerors dealt with them through institutions whose origins were Hispanic and indigenous, in such fields as labour organization and tribute payment. Historically, too, attempts were made to incorporate Amerindians into the Catholic belief system, and into European-oriented economies. For example, Amerindian labour tribute in the Andes was adapted and utilized by the Spanish in their silver mining ventures, or in cloth production in the *obrajes* (see Ch. 2).

In areas where indigenous populations were relatively thinly dispersed and/or not under state control, the dynamics of population

Plate 6.1   An Afro-Ecuadorian couple (*Source*: D Preston)

interaction between colonizer and colonized were distinct. Such indigenous groups were found in the Spanish Caribbean islands, in the central forests of the Amazon basin, and on the grasslands of the Southern Cone, as well in the northern areas of what is now Mexico. Such indigenous groups were able to avoid incorporation into labour and commercial relations with the Spanish conquerors. However, they were also severely affected by European-introduced disease, which greatly reduced populations such as those in Cuba. Today, surviving groups are more likely to demonstrate a sense of independence from the nation-states which purport to run their affairs. Although many in the North have an image of Amazonian groups being 'noble savages', the extent of their political organization and their facility with modern telecommunications and technology (such as video recorders) belies this simplistic image, and reminds us that they are very much part of contemporary global culture.

The inability of Spanish colonial structures to incorporate these groups as labour into the emerging colonial economy necessitated the

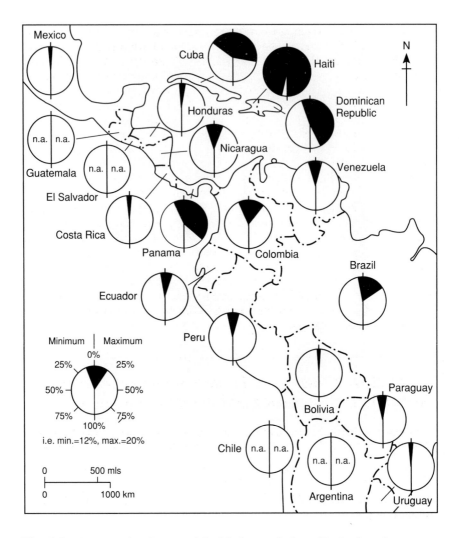

Fig. 6.1   Contrasted estimates of the black population of Latin America

substitution of alternative labour, and explains the presence of the third major population group in the continent – the African Latin Americans. Between 1451 and 1870, Spanish America and Brazil imported some 54 per cent of the African slaves coming into the Americas, a total of an estimated 5.12 million slaves; Brazil alone accounted for 31 per cent of the total. As argued by one writer, it was the plantation zones, selected for geographical and physical reasons, which became Afro-America, and black labourers were long associated with rural economies. Black identities in Latin America have become associated with syncretic religions with African antecedents, political organizations and myriad cultural forms in both rural and urban areas. As awareness of racial

150

discrimination has arisen, so black organizations have formed, such as the Brazilian Black Front in the 1930s. In regions with a large proportion of blacks, such as the Pacific lowlands of Panama, Colombia and Ecuador, blacks have created strong regional identities in response to discrimination and lack of historical recognition by governing (non-black) élites, and feel less part of a *mestizo* national community. However, in many areas, the processes of discrimination and *mestizaje* have led to the suppression of black roles in the creation of nations and development. In many countries, lighter mulattos identify as 'white', while blacks identify as 'mulattos' or '*mestizos*'. Only Haiti has adopted an affirmation of blackness, or '*négritude*', as a nationalist ideology.

The distribution of the black population of Latin America according to contrasted estimates (Fig. 6.1) shows, firstly, the degree of uncertainty about who was 'black' and also that the supposed blacks were highly concentrated in the Caribbean, Brazil and some countries bordering on the Caribbean (see Box 6.2 on the treatment of slaves). Within countries, black populations are likewise concentrated in certain areas. In Colombia and Ecuador blacks are largely confined to the Pacific coastlands and (in Colombia) the Caribbean coast. In Brazil, blacks and mixed-race people are widely distributed but concentrated on the northern and north-east coastal areas (Figure 6.2).

Box 6.2 Why was there different treatment of black slaves in Latin America and the United States?

One question which often interests students is why black slaves in Latin America and in the United States received different treatment. Whereas in Latin America intermarriage and tolerance led to the emergence of a continuum of racialized populations, involving indigenous and Asian populations as well as blacks and Europeans, in the United States by contrast the categories of white and black are mutually exclusive. The differences are attributed to five factors, which can be variously linked with the separate processes of manumission and miscegenation. In Latin America:

1. The sex imbalance in early settler populations led to high levels of miscegenation.
2. The manumission of offspring of black female slaves and white élite men, led to the emergence of a large, free, mixed population. (Similarly, children of indigenous women and European men were often freed from labour tribute obligations and were brought into the élite group.) It is also suggested that manumission, and even self-purchase, of slaves was more widespread in Latin America than elsewhere.
3. The Catholic church permitted the free marriage and bearing of children by slaves, where Iberian legislation recognized the unnaturalness of slavery.
4. The type of local labour markets and the presence or absence of indian labour enhanced the above processes.
5. Slaves in Latin America more frequently rebelled and escaped, especially in Brazil, leading to large, free, independent African Latin American groups. Slavery was abolished in most Latin American countries by the mid-nineteenth century, although Brazil did not liberate slaves until 1888.

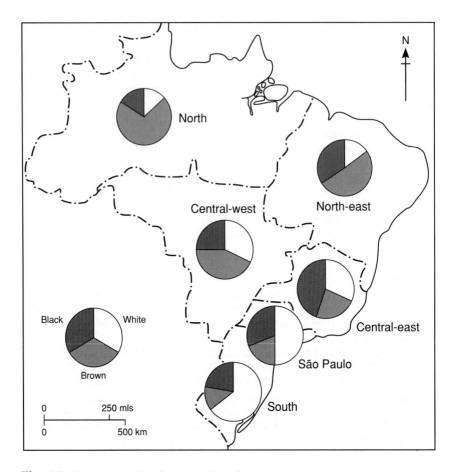

Fig. 6.2 Population distribution in Brazil

Incorporation of non-Hispanic groups into Hispanic culture and institutions was variable both in space and through time, yet indigenous and black groups generally persisted. Non-Hispanics lived by cultural norms and beliefs which are often a result of mixture, or *syncretism*, between Amerindian, black and Hispanic-Catholic beliefs. Their identities are dynamic and place-specific, affected as much by labour market participation and state policies as by supposedly timeless rituals. In Ecuador, for example, where indigenous peoples of the Andean and Amazonian regions carried out a nationwide uprising against the state in 1990, identities are shaped by Catholicism, the national education curriculum, by the demands on families' labour time, and by a state which has done little to respond to their demands for land.

European settlers in Latin America were few compared with other colonies around the world. Between 1509 and 1790, the numbers of potential settlers leaving Spain for the Americas reached only 150,000. Spanish and Portuguese immigrants were of course initially the major groups, but these were also specific in terms of gender and religion. Women immigrants were outnumbered by men for many of the early years of Spanish colonization, and for this reason, male Spaniards often partnered, violently or otherwise, with indigenous women, giving rise to the *mestizo* (mixed race) groups. Catholics were also the great majority, as early laws prevented the immigration of Jews or converts from Islam from entry into Spanish America. 'White' settlers adopted strategies to maintain what they saw as 'purity of blood', or descendence, which justified social hierarchies and patterns of exclusion. The control of women's marriages and premarital lives were a cornerstone in these class-enhancing strategies, and were largely succesful even when other European groups entered Latin America in larger numbers after independence. Italians, Germans, English, Welsh and other nationalities settled and in cases intermarried with existing élites. Although present in relatively small numbers, their arrival at certain ports and areas gave rise to spatially differentiated ethnic relations and regional cultures. Immigrants also tended to go to urban areas, rather than settle on the agricultural frontiers (although Welsh and Scots went to Patagonia, Mennonites to rural Paraguay, and Germans to southern Brazil and Chile, among others), and it is in the highly urbanized countries of the Southern Cone where their racial and cultural influence has been strongest. Argentina and Uruguay experienced substantial European immigration in the twentieth century, as labour was attracted into booming economies. Brazil, which attracted Italians to the coffee-growing regions of São Paulo, and Cuba, which experienced substantial Spanish immigration, were two other countries with major European influences in the current century.

Changes in the nineteenth century

Late in the nineteenth century, groups from the other side of the Pacific Ocean began arriving, often as labourers in the newly commercializing plantations and mines. Chinese and Japanese indentured labourers were brought over, and settled throughout Latin America, although the Japanese were particularly significant in Brazil. Unlike the African slave populations, the Chinese labouring immigrants were exclusively male, and consequently intermarried with each of the other racialized groups, adopting Catholicism and Spanish names in the process. Brought first into Peru and Cuba, the Chinese entered commerce after leaving manual work, and now are most numerically significant in Cuba, Mexico and Peru.

Immigrants from Japan arrived after the Chinese at the end of the nineteenth century, and Japanese Latin Americans are now found in Brazil, Peru, Colombia and Paraguay.

Gender relations

Gender relations in Latin America are often assumed to be based upon Latin American versions of Hispanic and Catholic notions of female *'marianismo'* (from María, the mother of Jesus Christ), and male *'machismo'* (from *macho*, or male). In other words, the model of gender relations is perceived to be based on female submissiveness to family authority, chastity, modesty and moral superiority, interacting with male displays of virility, aggression to other men, and control of female family members. These patterns of male and female behaviour and identity, based on concepts of honour and shame, were brought into the continent with the arrival of the Spaniards in the late fifteenth century, but have seen centuries of transformation. Not least was the racial element of white conquerors raping and marrying Amerindian women, which meant that power and a sense of status difference was strengthened.

However, in practice it is difficult to recognize these patterns at the present time, for four major reasons:

1. Firstly, distinct indigenous (and black) notions of gender ideology, and practices of gender interaction persist in many areas of Latin America. For example, in the Andean region, indigenous Quechua speakers of both sexes can have premarital sex, can have land use rights in the villages, and participate in communal work parties to cultivate land. This is not to say, however, that gender inequalities do not exist: the absence of the *marianismo–machismo* model does not mean the absence of practices which limit women's choices and opportunities for development. Girl children are less likely than their brothers to be sent to school in the Andean communities, while village politics silence them and do not take their de-velopment priorities into account.
2. Secondly, class differences between groups have, throughout history, transformed gender relations. Although the élite classes have had the wealth and security to maintain women in the home, poorer groups have depended upon the labour force of each family member. As a result, low-income families have seen labour force participation of women and men, young and old, in their attempts to combine sufficient income from various sources to survive. In rural areas, women and men participate in various agricultural activities, depending upon the specific sexual division of tasks which varies geographically. With increasing diversification of economies and with urbanization, the opportunities for women and men have increased; male and female migration into larger cities reflects this.
3. Thirdly, the rising number of female-headed households in the continent suggests that women's and men's roles within the family are being transformed. It is estimated that some 20 per cent of households are now effectively headed by women. Women are

raising children on their own, and getting by through working to earn money and by sending their children out to work. Rather than continuing in a violent or unsatisfactory relationship, the evidence suggests that certain women are leaving traditional nuclear families to live independently. In many cases, this entails a fall in total income and its security; it can also be associated with the feminization of poverty in many areas of Latin America.

4.   Fourthly, political and social struggles have changed the legal and ideological context for male–female relations, away from those of the past. In fields as varied as divorce, contraceptive use and state anti-discrimination legislation there have been attempts to give women more choice, independently of their families. While in many countries it used to be a legal requirement that women request their husbands' permission to work, nowadays women choose more freely whether to work or not. Feminist movements in the 1970s pushed for the abolition or transformation of discriminatory laws and social practices, and have had a lasting effect on women's personal identities as well as their legal position.

Currently, gender differences in the labour markets are marked in most countries of Latin America. Throughout Latin America, women made up around 20 per cent of the total labour force in 1950–70, but it would appear that it has risen to around 26 per cent with economic crisis and the pervasive impacts of structural adjustment programmes (see Boxes 6.3 and 6.4). In both Ecuador and Mexico we can see profound changes in everyday life, economic and social changes which act upon the social actors shaped simultaneously by gender, generation, location and class.

Box 6.3   Women's work under conditions of economic crisis: Ecuador

In Ecuador between 1982 and 1988, eight distinct structural adjustment programmes (SAPs) were applied to an economy which was increasingly facing crisis. With far-reaching effects on low-income groups generally, the SAPs are also recognized to have gender-differentiated effects. In one low-income neighbourhood in the industrial port city of Guayaquil, men and women have been affected in various ways. Overall, households have had to send more members out to work, with a jump from 19 to 32 per cent of households with more than three working members. Between 1978 and 1988, men's work has become more insecure, with some jobs created in the export-oriented shrimp industry, but otherwise contraction in the building small crafts and artisanal sectors. Women's labour participation increased from 40 to 52 per cent over the period (well above the Latin American average noted above), while those in work were working longer hours. For domestic servants, cooks and washerwomen, as well as sellers, 12–18 hour days became the norm, sometimes for six days per week. Generational effects can also be seen: daughters were increasingly taking over reproductive tasks from mothers, and reducing time spent in education. Eldest daughters took on childcare at a young age, and started cooking for the family at 10 or 11 years.

Source: Moser (1993).

Box 6.4   Women and Economic crisis: Mexico

In Mexico, similar changes are occurring in low-income households. Here, the profound economic crisis from the late 1980s has had severe consequences for the organization of labour in poor families. A study in the city of Guadalajara found that women were key actors in the new survival strategies. Family sizes increased between 1982 and 1985, as did the number of workers per household. Two groups have particularly been affected by the move of family members into waged work: women over 15 years, and boys under 15 years. Boys (and some girls) under the age of 15 years left education and entered the labour market, or took part-time jobs enabling them to continue in education. By contrast, young women under 15 years played a growing role in domestic organization, freeing older women for waged work.

Source: González de la Rocha (1988).

Moreover, the involvement of women (and children) in various combinations of reproductive and income-generating work means that statistics on labour force participation are skewed, if not inaccurate. Nevertheless, it is possible to outline the major fields of gender-segregated work. Domestic service is a major gender-segregated sector for women, with about a quarter of Latin America's working women here; for example, 25.5 per cent of Bogotá's female labour force and 32 per cent of Brazil's. In newly developing globalized production sectors, women also have been recruited disproportionately, although the numbers of women employed in export strawberry and flower production, export clothing production and so on are relatively small. Located in specific locations around Latin America, such as along the US–Mexico border, around Bogotá and in northern Ecuador, these industries rely upon non-unionized, generally young, often migrant, women.

Control over sexuality and bodies is one area where gender differentials are also marked. Although in many countries, there is a separation of the church and the state, Catholic attitudes to birth control have restricted acceptance and use of these. Women's access to abortion in Latin America is either illegal or restricted (permitted solely in cases of rape, incest or to save the woman's life), except in Cuba where it is available on broad socio-medical grounds (see Box 6.5). In the rest of Latin America, many women are thus forced into unhygienic and unsafe conditions for illegal abortions: it has been estimated that 30–50 per cent of maternal deaths are a result of poorly performed abortions.

Generational and ethnic differences also become a factor in body politics in Latin America. As young women are not generally expected to become sexually active, there is little sex education. Also cases of rape of young women are hard to prove, leaving young women unable to have legal abortions. In São Paulo, Brazil, it is estimated that nearly 15 per cent of the 14,000 pregnant girls aged 11–14-years-old each year have been raped, yet they have little recourse to hospital treatment.

Box 6.5 Abortion in Brazil

Although abortion is a criminal offence and condemned by the Catholic church, it is widely practised, and an estimated 4 million a year are carried out. Only abortions to save a mother's life or in proven cases of rape are permitted, leaving many women to go to backstreet practitioners. As a consequence, up to a third of hospital beds in Rio de Janeiro's obstetrics wards are taken by women suffering the after-effects of illegal abortions. Women's groups wish to see a decriminalization of abortion, and are attempting to raise the issue in a country where such debates are not on the public agenda.

Among indigenous populations, there are local birth control methods, yet in indigenous villages of Bolivia and Peru there have been forced sterilization programmes carried out by governments without women's consent, and in Brazil some black women have also been subject to forced sterilization.

Generation: young and old in Latin America

Children, adolescents, adults and older people all face different situations in contemporary Latin America. While many have a family role, factors such as labour market participation, poverty, war, migration, family disintegration and abandonment influence the benefits and costs of development experienced by individuals at different stages of their life cycle (see Table 6.1).

Children often play an important role in family income-generation, particularly among peasant and low-income urban populations, as their labour either generates income directly or substitutes for adult labour in the household. In rural economies, children participate in animal herding, water and firewood collection, harvesting and care of younger siblings by the time they are 10 years old, and in certain circumstances migrate to urban or rural labour markets soon after. In Latin America as a whole agriculture is the sector which most absorbs child and adolescent labour (a total of some 4.6 million, or 50 per cent of those aged under 20 years), due to the large proportion of family farms. In urban areas, too, children undertake work for families, combining in some cases a half-day at school with work. Shoe-shining, selling, running errands are some of the income-generating tasks done by children, while unpaid tasks such as childcare, cooking, cleaning and collecting water are others. The service sector, including domestic service, is, alongside agriculture, a significant employer of children and young people. Because of participation in family labour and the informal sector, estimates of the numbers of children working in Latin

Table 6.1 Generations: birth, death and education.

	IMR[1]	Life expectancy[2]	% of schooling primary enrolment[3] Boys	Girls	% literacy[4] Men	Women
Bolivia	102	55	88	78	85	71
Peru	82	63	–	–	92	79
Guatemala	54	63	–	–	63	47
El Salvador	59	64	71	62	76	70
Honduras	63	65	89	94	76	71
Ecuador	60	66	–	–	88	84
Brazil	60	66	–	–	83	80
Nicaragua	56	65	74	79	–	–
Dominican Republic	61	67	–	78	85	82
Paraguay	41	67	91	84	92	88
Colombia	39	69	72	74	88	86
Mexico	40	70	–	–	90	85
Venezuela	35	70	107	88	87	90
Argentina	31	71	–	–	96	95
Panama	22	72	90	89	88	88
Chile	20	72	–	–	94	93
Uruguay	22	72	–	–	97	96
Costa Rica	18	75	85	85	93	93
Cuba	11	75	97	94	95	93

[1] Infant mortality rate, per thousand, in 1990
[2] In 1990, at birth
[3] Net figures for 1986–89
[4] In 1990

Source: UNICEF (1992) Tables 1,4.

America vary, but reached 9.3 million children aged under 20 years, in 1970 (see Box 6.6). Since that period, the impact of structural adjustment programmes and the 'lost decade' of the 1980s has probably increased the number of working children. For example in Lima, Peru, children aged 10–19 years formed 9.7 per cent of the formally registered economically active population in 1982.

Few countries have legal safeguards for children. Ecuador was the first country to sign and ratify the international Convention on the Rights of the Child, and Brazil in 1990 implemented a statute for constitutional safeguards for the rights of children and adolescents. However, in practice, the experiences of children depend more on class, race and family situation, than on legal provisions. Class determines the extent to which children can enjoy a childhood, a concept which only became widespread in Latin America in the nineteenth century. Wealthier children are more likely to survive birth and infancy, to continue in full-time education into adolescence, and to be well nourished. The exception to this pattern is Cuba, where childhood has been the recognized time for education into citizenship and work, and

---

Box 6.6 Street children in Latin America

Although children live on the street in the big cities in every Latin American country, there are certain factors which increase the likelihood of them living on the street. Large income differentials, lack of support institutions, civil disturbances, war and migration flows all contribute to the presence of children who live varying proportions of their lives on the streets of cities. By living (often in a group) on the streets, they obtain freedom from violence and persecution at home and find comfort and support with others in a similar situation. But they are vulnerable in ways that other children are not. They exchange victimization from a step-father for victimization and repression by the police. In Colombia, young street people have become workers for the drug mafias. In Duque de Caxais, Brazil, between one and four street children are killed each week by the police, drug barons or by gangs, in order to 'clean up the town'. In a society where blacks are generally poorer and have fewer opportunities, the 'cleaning of the city' is racialized. In the state of Rio, Brazil, in 1992, 470 street children were killed; 80 per cent of them were black in a 60 per cent black area.

---

where life expectancy and literacy rates are high and child mortality low.

Survival rates in infancy also depend on geographical location and race, with higher mortality rates among indigenous and rural groups. Studies in Bolivia found the highest death rates among rural highland villages (infant mortality rates, IMR, of 223/1,000), but interestingly found higher rates of infant death in urban highland areas (IMR 157/1,000) than in rural lowland locations (IMR 140/1,000). In rural Andean areas of Bolivia, indigenous groups suffer still higher rates of infant mortality, owing to social isolation and lack of adequate health care. Some 70 per cent of infant deaths go unreported, especially girls', suggesting that ideologies result in greater care of boy infants.

Children's priorities for development are in some cases distinct from those of adults. In a poll carried out in Costa Rica and Ecuador, children aged 5–14 years chose protection from drugs, sex abuse, and all forms of violence as their priority. Such a concern reflects the vulnerability of children to the international drugs trade, to power differentials between adults and children, and to civil disturbances. In Peru, the war between the Shining Path and the military has had severe consequences for children. As the war has extended throughout the country, young people have been displaced from their homes, orphaned and, owing to the worsening economic situation, have undertaken paid work rather than continue their schooling. In Lima, Peru, school attendance figures in 1991 were down 50 per cent, while the government reckons that some 2 million children aged 6–14 years are working, despite legal barriers. Other children, some as young as 8-years-old, have been recruited by Shining Path, while the military increasingly targets young people in the war.

At the other end of the life cycle, Latin Americans face a different situation and priorities. Moreover, there are relatively few data on the older populations in Latin America. Again as with children and adolescents, class, race and family structure as well as geographical location and gender differentiate among older populations in Latin America, making it very difficult to generalize about this group across the continent as a whole. One factor which varies spatially is the 'age for ageing', that is the age at which work is expected to fall off and physical deterioration begins. For example, the age for ageing varies from about 30 years in Potosí, due to socioeconomic and physical factors, to around 75 years in wealthier urban areas of Mexico. Regional differences within countries affect life expectancy, and hence experiences of old age. In north-eastern Brazil, life expectancy is eight years less than the national average, while in Guatemala and Honduras there is lower life expectancy in villages than nationally. Among indigenous communities in the Andean countries life expectancy is lower than in urban and *mestizo* populations, owing to poverty and social isolation (see Box 6.7).

Box 6.7   Effects on old lives in Latin America

A recent study points to various social, geographical and political effects on elderly populations, which vary from one area to another. These groups of effects are named according to one typical area where they occur:

1. Vilcabamba (Ecuador) effect: extra longevity, retirement unknown.
2. Potosí (Bolivia) effect: low survival rates, early disability.
3. Beni (Bolivia) effect: remote, poor services.
4. Bogotá (Colombia) effect: metropolitan social services, although not for massive need.
5. Lima (Peru) effect: migration to city, geographical segregation, some abandonment of old.
6. Belize effect: small country, emigration, abandonment of old.

Source: Tout (1989).

Development priorities for older people include such factors as security of income and/or support, social interaction and appropriate health care. However, for many Latin American elderly populations these needs are not recognized by nation-states and society in general, thereby marginalizing them and reducing their contribution to society. Security of income is problematic for the elderly, who, as their physical strength diminishes, are less valued as workers, especially in manual labour. In Brazil, workers over 40 years of age expect to be discriminated against, both in agricultural labour and in qualified and semi-skilled work. One Brazilian elderly person explains the social marginalization this entails, 'When you leave off producing, you lie around like an outcast till you die.' For older women particularly, labouring work becomes more difficult, although their reproductive work continues to be useful. Older women often head female-headed households in low-income groups, in which they carry out cooking and childcare tasks, while adult daughters do wage work. Development programmes have recently attempted to provide alternative income sources for the older populations. From a bakery run by the elderly in Bogotá, Colombia, to a crèche run by elderly women in Brazil, these programmes are mostly urban, although older rural populations have few income sources. Income sources for the elderly in the form of pensions are limited throughout Latin America. Formal social security cover varies from 80 per cent in Costa Rica to 1.5 per cent in Haiti. Richer countries such as Argentina and Costa Rica have been better able to provide pensions, although unequal provision for different workers (and women) remains.

Social interaction is valued highly by elderly people, as their working roles diminish. However, in Latin America as elsewhere, extended family networks are becoming rarer, leading to the marginalization and in some cases, abandonment of older kin. Elderly relatives are more frequently abandoned in the 1990s, although this was rare in the 1950s. One survey in Peru found that 58 per cent of the elderly had no place in an extended family. Rural–urban migration which transformed the demographic make-up of Latin American countries in the decades between 1940 and 1980 (see Ch. 7) also meant that older people were left behind in rural areas, with few kin and support services. Housing in urban areas is more oriented towards nuclear families, thereby putting strain on limited state provision of alternative housing. Old people's homes are more common in Latin America than in other less-developed countries. However, owing to inadequate budgets and undertrained staff, these homes are not generally welcomed.

The breakdown of extended families affects men more than women, as the former are seen as more of a burden once their productive life has finished. Nevertheless, as women outnumber men in the older age cohorts, women also make up the bulk of the elderly population. Elderly women are also disproportionately located in urban areas. Urban sex ratios for people of 55 years and over show uniformly the numerical dominance of females in Latin America. In urban areas of Uruguay for example, there are only 74 men over 55 years to each 100 women in that age group. Gender differentiated roles and ideologies make the roles of older women distinct. For example with ageing, women become less subject to male authority, although they can also be more vulnerable and lonely. An average of between 35 per cent (Mexico, Cuba) and 50 per cent (Brazil) of women over 60 years are widowed; compared with rates of between 7 and 16 per cent for men. In Mexico, it has been claimed that older women would rather be a domestic servant than be alone once widowed, while in Colombia grandmothers are more likely to be cared for by grandchildren.

One other signficant priority for older people is appropriate health care. However, nation-states in Latin America have not made provision a major goal. Poorer countries such as Peru, Ecuador and Bolivia have low levels of hospital provision, whereas Argentina offers a government programme providing integrated medical attention for nearly half of the over-60s.

In some Latin American countries, older people have organized to demand resources and attention from government and society. In Uruguay, 160,000 older men and women have an organization which deals directly with ministers of government. In Argentina, the Grandmothers of the Plaza de Mayo organized themselves to find their children and grandchildren who had been 'disappeared' by the military junta in the 1980s. In this case, women felt that their 'power of having no power' (due to gender, age and lack of institutional positions) allowed them to accomplish things that younger women and men could not.

Race, gender and generation: the case of domestic service

The position of younger women employed in the domestic service sector highlights their relative powerlessness, as well as the simultaneous influences of gender, race, and generation in the constitution of everyday lives in Latin America.

In the Andean countries, where impoverished indigenous peasants provide labour for Spanish-speaking, *mestizo* urban centres, domestic service is a major occupation for young women. Single women, most of them under 20 years old, migrate into cities and towns to work for middle class and élite families. Working and living in their employers' homes, they clean, cook, care for children and do daily shopping. The fact that these domestic employees are generally young, indigenous and female means that they are at the bottom of several social hierarchies, and are generally treated poorly. As they receive low wages, few employment benefits or security, and have generally few opportunities for education, their opportunities for socioeconomic advancement tend to be limited. The creation of a labour market in domestic service, segregated by gender, race and generation, has also created its own spatial quality, a pattern of life which is lived through, and defined by, its geographical location and range. For example, *'empleadas'* usually have restricted daily mobility, confined largely to the employer's home. When outside the kitchen and yard, *empleadas* are often 'invisible' to the employers' families. Except for visits to markets and shops, social visits are limited to one day off a week on average. Domestics are also subject to verbal and occasionally sexual, abuse from their employers' families. One domestic employee explains,

> I thought she loved me because she told me: 'I love you like a daughter; here you have everything.'

The example of domestic service highlights how the spatial and social relations based on race, gender and generation simultaneously and inextricably locate individuals in Latin American societies. Although indigenous groups are not present in every region of Latin

America, domestic service in these areas can also demonstrate racialized social relationships. While in the central Andean region domestic service comprises a segregated labour market for young indigenous women, in areas of Colombia for example it is young black women who fill this occupational niche, and experience similar disadvantages. The often difficult relationships between *empleadas* (domestic servants) and *patronas* (female employers) highlight the difficulties of assuming a common female identity, across lines of race, class and generation. This factor is recognized by the organizations of domestic servants in Latin America, who criticize middle class feminist organizations for continuing to use domestic servants, and highlight the problems of assuming a common feminine experience.

Conclusions

The three dimensions of social differentiation examined in the chapter, although considered separately, together constitute and perpetuate social and economic relations between Latin American citizens. These aspects of individual and social identities continue to underlie the use of space and resources by groups. At spatial scales as diverse as the household, labour markets, cities and the nation-state, participation in development varies with race, generation and gender. Types of discrimination are associated with each of these social markers, and

Plate 6.2   The juxtaposition of different cultural expressions is characteristic of Latin America, as in this scene from Cusco, Peru
(*Source*: S Radcliffe)

limit opportunities for children and the elderly, for indigenous and black groups, and for many women, whose priorities for development are not fully taken into account by the decision makers and managers of national development plans.

Further reading

Cubbitt T (1988) *Latin American Society*. Longman Scientific and Technical, Harlow. (Includes chapters on gender, ethnic and class relations.)
Green D (1991) *Faces of Latin America*. Latin American Bureau, London. (Includes chapters on indigenous people, women's work and cultural change.)
Jelin E (ed.) (1990) *Women and Social Change in Latin America*. Zed, London. (A collection of country-based and theoretical studies.)
NACLA (1992) 'The Black Americas: 1492–1992', *NACLA Report on the Americas* **XXV** (4). (An historical and contemporary overview.)
Winant H (1992) 'Rethinking race in Brazil' *Journal of Latin American Studies* **24** (1), 173–92. (Uses Brazilian material to suggest a new approach to racial issues in Latin America.)

CHAPTER 7

People on the move: migrations past and present

David Preston

Introduction

Migration (see Box 7.1) is a part of everyone's life cycle. Children migrate with their parents, or on their own to be educated, on marriage and when seeking a living. Population movements have always been part of Latin American urban and rural life: there is little evidence that overall mobility changed greatly either during the colonial period or in the present century. What merits special attention is the composition and direction of migration. Migration at any given moment mirrors stresses in households, in society in general and in different regions.

Box 7.1 What is migration?

Here we mean migration to refer to any change of residence that takes place for more than a few days, whether or not the migrant is accompanied by other members of the household. Census definitions such as habitual residence may alternatively be used and it is important to realize that for every change in definition, a different measure of human mobility is being taken. Another term frequently used in migration studies – permanent migration – deserves elaboration. Migration permanency is a very elusive concept. Long-term Mexican migrants to Chicago may refer to their migration as temporary even after 25 years' residence because they perceive their home village in Mexico as the place to which they will ultimately return. If they die in Chicago, can their migration to the USA then be seen as permanent because of the accident of their death there? Some forms of migration – journey to work, holiday, marketing – may be cyclic and thus clearly temporary but the use of a word to qualify migration, which refers to probability of future mobility, is hazardous and thus best avoided.

In this chapter migration is referred to both as a historical process and in rural and urban worlds which implies a neat dichotomy between places which have clearly recognizable characteristics. Rural places contain people and households whose livelihood is largely derived from farming, while urban people are engaged in manufacturing, the distribution of goods and the provision of services. Rural people live in an environment in which households live in small agglomerations of maybe 10–100 dwellings or in dispersed homesteads; urban people live in an environment comprising closely packed dwellings and other buildings where many thousands of households live in close proximity.

Such a neat division of human environments is inadequate with respect to migration simply because human mobility is such that urbanites seek rural residences, while some farmers choose to live in towns. Many households are involved in different ways of getting a living which encompass urban and rural environments for different periods of time and at different seasons of the year. Many Latin American rural families seek to urbanize their environment even though the whole household is predominantly occupied in farming and farming-related tasks. It is therefore unrealistic to suggest that people can be assigned to two mutually exclusive categories – urban and rural. Thus a division of mobility into that experienced by rural and urban people is inherently artificial and merely a device by which a series of mobility patterns can be described to enable some conclusions to be drawn about the extent to which mobility does provide an insight into the ways people use different environmental resources and into the pressures on people from the political, social and economic environment of which they are part.

Migration theory and Latin American experience

The examination of the state of knowledge about migration in Latin America must make reference to the substantial body of theory that exists regarding population mobility and of the alternative theoretical frameworks within which migration can be discussed. Much of migration theory is based upon emphasis on the economic motives underlying migration which are most readily quantifiable and which do provide a powerful if partial explanation of migration. People move towards areas where wage levels are high (cities) and where quality of life, expressed by housing availability, and the degree of provision of health and education are most satisfactory. Similarly, as we have already suggested, migration between countries is predominantly from poor countries towards richer countries.

There is also a group of migration theories which state that most migrants move short distances – in part a response to the Principle of Least Effort (This principle, phrased in relation to migration, states that people will move most frequently over short distances.) – and that the larger size of a populated centre the greater the attraction that it exerts on migrants. Important fundamental work on migration has been carried out in Sweden by Hägerstrand (1957) and more recently by Carlstein (1982), which demonstrates the patterns of mobility associated with different phases of the human life cycle and with different roles of individuals within households.

Care is needed in recognizing the level of explanation that is being sought. When a migratory event occurs, the causes are complex and cannot be subsumed by referring to abstract situations such as 'population pressure' or even soil erosion. It is necessary to know the whole context in which migration is occurring, the extent to which the decision is made by the individual or the group and the length of time over which the decision has been considered. When interviewing a migrant she may explain that she moved to the town after a big row with her husband; but after talking further it emerges that she moved also to enable her daughter to attend a better secondary school, that she moved to live with an aunt who needed help in her shop, and that her two cows had died of foot-and-mouth disease and she had no alternative form of income. Thus a bundle of possible explanations of migration emerge, without it being clear which, if any, can be identified as the 'real' cause.

The high-level changes that are occurring in national and local societies may also be related to migration and various writers have pointed out that it is in the interest of larger-scale industry that there should be a plentiful supply of labour, to help keep wages down and to facilitate the maintenance of a rapid turnover of labour in response to a changing economic situation. The move from labour-intensive to capital-intensive industry – including agriculture – also affects the need for people to move about to seek a living. Thus any analysis of migration should take into account the changes in society at large as well as those changes at a local, community and household level.

In this analysis of population mobility in Latin America we shall seek to report the situations in which migration is taking place and emphasize the impact of migration on both sending and receiving areas. We shall also try to show the extent to which migration is part of the human geography of rural and urban Latin America and the degree to which what is happening there accords with existing theory on human mobility. Migration is nothing new, and the history of human movements to Latin America summarizes the historical human geography of the region.

Migration in the past

Nineteenth-century immigration

By the time that independence movements were growing in importance in the early nineteenth century, the frontiers of settlement in eastern and southern Latin America were still relatively close to the coast. In Argentina Indians presented a serious drawback to settlement in the pampas until the last part of the century, even though there were few new settlers. In Brazil mineral discoveries (gold and diamonds) had led to settlement in the uplands of Minas Gerais by 1700, and a shift of movement both westwards and southwards combined with a wave of Portuguese immigrants in the last half of the eighteenth century were partly the cause of a ten-fold population increase during the eighteenth century. Most of the migrants were men and widespread miscegenation resulted rather than the creation of a cultural enclave. But in 1800 São Paulo was still a frontier town belonging to the *bandeirantes,* armed bands of wandering slavers and traders, even though coffee was being planted in the Paraiba valley to the north using slave labour for export through Rio. The southern part of Brazil, like the west, was sparsely populated.

During the nineteenth century, the most important changes in the population of Latin America were not associated so much with the independence from Spain and Portugal but with the wave of European immigration that peopled some of the emptier areas in the Southern Cone (Chile, Argentina, Uruguay and southern Brazil); this provided a commercial and social stimulus as foreigners began to farm parts of the lowlands and to develop commerce, transportation and mining in other areas. Spain's commercial hegemony had long since waned and Great Britain, France and Holland became major traders for the continent as they sought markets for the goods and machines that the Industrial Revolution produced. The isolated areas of the empires – the south in Brazil and the River Plate in Spanish America – had long suffered neglect and restrictions on free trade and the severing of these controls increased the range of contacts that such areas made with the northern hemisphere. While the development of coffee and bananas as export crops in Central America in the second half of the nineteenth century was the result primarily of foreign investment and immigration, the latter was small scale since the labour was locally available: Indian lands were being appropriated by non-Indian townsfolk forcing the Indians to seek at least seasonal employment elsewhere.

While the abolition of slavery was also a powerful factor stimulating the development of new social relations with production, only in a few areas was the labour shortage solved by importing indentured labour. In Peru, for instance, in the middle of the nineteenth century, Chinese and later Japanese were brought in large numbers to provide labour for cotton and sugar plantations on the coast (Faron 1967). Chinese were likewise imported into Mexico, Cuba, British Caribbean islands and Guyana. They came from humble backgrounds but, as the size of the Chinese (and Japanese) community grew, other Chinese migrants arrived with greater economic resources in order to profit from trade with their home community. In Cuba the existence of empty areas with a good potential for farming was attractive and, even before Independence, Spanish-speaking migrants came there from Hispaniola (following a black rebellion) and Louisiana (when it was transferred to France in 1803). Migrants from France founded Cienfuegos in 1819 and many refugees from the civil wars attendant on independence found refuge in Cuba. In addition, large numbers of Africans were imported into Cuba until the late 1860s despite growing efforts to end the slave trade.

The Southern Cone saw the most important transformation as a consequence of an increasing volume of immigration during the nineteenth century but this swelled to a flood during the last quarter of the century which continued undiminished until the start of the First World War in 1914. This phase of immigration was distinctive because it effectively settled a large temperate part of the continent with a predominantly European population. This settlement was closely related to the creation of new means of communication between Europe and the New World – steamships – and overland – mainly railways – in Argentina especially, with the use of foreign, mainly British, capital. In the southern part of Brazil colonization by Europeans was much less tied to railway or road building although in the coffee zone of São Paulo railways were just as important as in Argentina and eventually became major reasons for intensification of beef production beyond the Parana river.

Immigration into São Paulo state was closely associated with coffee production and in the 1880s Italians replaced Portuguese as the most numerous group (62 per cent), many of whom were subsidized migrants destined for the coffee zone. Most came from northern Italy (69 per cent in 1876–1900) at a time when southern Italians were predominantly migrating to the USA (Merrick and Graham 1979) and São Paulo planter organizations actively sought migrants from northern Italy. Many of the Italian migrants came as whole households and the access to land, even as tenants, with the opportunity for various members of the household to earn wages at least seasonally, allowed many to become landowners or to enter business in the cities with some capital resources. By 1920 immigrants owned 27 per cent of all rural properties (and 19 per cent of the total property value) in the state of São Paulo; by 1932 immigrants owned 40 per cent of all the coffee trees in the state and the Italians alone owned 22 per cent (Merrick and Graham 1979). Further south in Brazil, immigration was rather different; it was smaller scale and from a wider range of sources with a proportion of the immigration being spontaneous although state governments and private colonization companies were also important organizers or stimulants to foreign immigration. German settlement predominated at first and São Leopoldo, the first European colony, was founded in 1824 just north of Porto Alegre in the southernmost state of Rio Grande do Sul. After 1870 Italian colonies were established and in the last quarter of the century German, Italian and Brazilian colonies had settled a good deal of the forested tablelands in the eastern part of Rio Grande do Sul. In Santa Catarina, further north, another important German colony was established near the coast and by the 1890s settlers were establishing themselves further inland. During the same period the newly created state of Parana was being settled by Germans, Italians, Poles and Ukranians stimulated by the construction of railways linking the newly cleared lands with the coast and with São Paulo.

In the river Plate countries independence early in the nineteenth century similarly created an awareness of the need to populate the country. The confirmed threat to farming presented by Indians in the areas west of Buenos Aires discouraged settlement here until after their extermination in the Conquest of the Desert (1879–83) but in Santa Fé and Córdoba provinces immigration was encouraged to settle the land west of Rosario. Colonies were established along the Rosario–Córdoba railway by the Central Argentina Land Company in the 1870s and private settlement schemes also existed. New settlers moved the frontier northwards in the 1880s into Entre Ríos but further settlement on a large scale did not take place until after the First World War when quebracho trees were exploited and settlers came from Paraguay in the north and from German and Slav areas of Europe.

The immigrants were not all farmers. They came from every part of Europe and included the nobility and the poor. Argentina, which had a population of 1.2 million in 1860, received 2.5 million immigrants in the next 50 years and by 1910 three out of every four adults in Buenos Aires was European-born. From the south of Italy came the *golondrinas* (swallows) for the wheat harvest, single men who returned home after the end of the southern hemisphere harvest to work their own fields. Many migrants remained, some stayed in the rural areas of the pampas attracted by tenancies which offered short-term security and the possibility of saving money. Others stayed in the cities, in particular Buenos Aires. Eighty per cent of the immigrants were Italian, and while southern Italians were predominantly attracted to rural areas, the northerners stayed largely in urban centres.

In Uruguay, likewise, immigration facilitated the growth of the farm (and national) economy but the proportionate importance of immigrants was less than in Argentina which in 1914 had 30 per cent of its population born overseas compared with 17 per cent in Uruguay (in 1908). The foreign-born population (mainly Italian and Spanish) tended to settle close to the river Plate and the river Uruguay (Crossley 1983).

Twentieth-century immigration

The great tide of European immigration to Latin America that peopled the southern countries grew in the last quarter of the nineteenth century. In the twentieth century the same processes continued as land further from the main cities and agricultural regions was settled. In Brazil settlers reached the Mato Grosso and the margins of the Amazon basin and the settlement of the grasslands began in earnest. In Argentina settlement extended further north towards Paraguay in the Chaco, Misiones and Formosa where German-speaking migrants from eastern Europe came in the 1920s and 1930s. Irrigation works in northern Patagonia in the Rio Negro valley and the arrival of the railway permitted more settlement and the extension of intensive agriculture further north in the Mendoza area created further opportunities in the early part of the century for migrants, many of whom were from Europe.

Elsewhere in Latin America small numbers of immigrants were establishing themselves as the cities grew and commercially oriented farming, sometimes organized by foreign companies (such as Grace in Peru, or the United Fruit Company in Central America), in lowland areas became more important. In some cases particular groups of immigrants such as the Germans, became associated with a single crop, such as coffee in Nicaragua or sugar in Cuba, but usually their total contribution to production was small even if they produced a disproportionate amount of the best quality goods.

Refugees of the 1930s–1950s

An important, yet seldom discussed, characteristic of twentieth-century immigration to Latin America is the diversity of migrant origins, and small numbers of settlers from a particular cultural group have often made a very distinctive contribution to regional and national life. The most striking of these groups includes the various German-speaking migrants of the twentieth century. This includes Jews expelled by Hitler Germany, Sudeten Germans and Nazis and Nazi collaborators fleeing the consequences of the end of the Second World War. Many were relatively wealthy business people and skilled craftspeople and they found it possible to insert themselves into Latin American society and to achieve economic success. In the 1960s, but to a lesser extent in the 1980s, many of the better hotels, photographic businesses, and money exchanges were owned and directly run by German-speaking people. Thus, when seeking a good small hotel in the Bolivian Yungas, a photographer selling good-quality postcards or a source of dry Spanish sherry, the chances were that you would deal with a German-speaking immigrant.

Lebanese immigrants

A second small but important group of migrants similarly encountered throughout Latin America are the *turcos*. The term covers Levantine people, many of whom came from Lebanon and elsewhere in the eastern Mediterranean after the First World War. Like the Germans the Levantines were victims of political upheaval and comprised middle-class families with the capital to get to Latin America and to start businesses once arrived. Almost exclusively urban, they established shops, typically selling a wide variety of products appealing to a range of customers, but others became established in industry and in financial enterprises.

East Asian immigrants

Oriental migrants are highly visible in Latin America even though they are not numerous. Asians were brought to Latin America since the decline of the economic role of African slaves in the nineteenth century. Indentured labourers from south and east Asia were brought to areas of plantations to fill the gap left by the departing slaves. In parts of the Caribbean and the Guyanas the numbers of indentured southern Asians in territories with only a small total population meant that they formed in time a substantial proportion of the population and remain today as a sizeable and visually identifiable group in such areas. Asians are a powerful political force in both Trinidad and Guyana. Elsewhere in

Latin America East Asians are the most commonly identifiable Asian group. Many came during the second half of the nineteenth century, both from Japan and China, and, although many came as contracted agricultural labourers and a proportion returned, others acquired the means to improve their status, to work as sharecroppers, purchase smallholdings but most commonly to return to the towns and to start small businesses without any link with their farming experience. The experience of the Japanese in the Chancay valley north of Lima differed from that of the Chinese. They became deeply involved with the spread of cotton-growing, as sharecroppers, managers and as organizers of commercial enterprises from the first decades of the twentieth century. Anti-Japanese feeling during the Second World War provided a convenient excuse, as it did elsewhere in the Americas, for expropriation of many Japanese assets and for a decline in their importance in commerce and agriculture.

Elsewhere in Latin America, Asians have been important elements in the population that has settled new lands, especially in Brazil and Bolivia. The numbers have always been relatively small but groups of East Asians, both Chinese and Japanese, have often been seen as relatively more successful than Latin American migrants. There is only limited evidence to support this conclusion, but the cohesive and well-organized nature of such culturally alien groups often gives them advantages over individualistic Latin American colonists receiving only limited guidance from government agencies. The most distinctive long-term contribution of these descendants of international migrants is the businesses which they now carry on, the most visible of which is catering. 'Chinese' restaurants are a feature of the urban scene in many parts of Latin America where Asians came in search of a better living than at home, but in other sectors of the business world too Chinese and Japanese have local and sometimes national importance.

Havens for the persecuted and pursued

The underpopulated areas of Latin America have long been a haven for groups suffering persecution in their homeland because of their beliefs. The best known group is the Mennonite religious group who migrated from eastern Europe to Canada, and some then moved to Mexico and to Paraguay, as well as to Bolivia, Belize and elsewhere. Seeking freedom to worship freely and strongly independent, they formed successful pioneer groups and are now well-established providers of dairy produce (Belize), cotton or beef (Paraguay). The Welsh too sought freedom, both political and spiritual, in their settlement of part of Argentinian Patagonia in the late nineteenth century. In the 1970s and 1980s white South Africans moved to farm relatively empty areas of southern South America, just as those with riches from drug trafficking in a neighbouring country or fleeing from the forces of justice in Europe sought refuge in cities and rural areas where investment could provide a good return.

Each of these groups shares the common characteristic that their limited numbers and common cultural and geographical origin made collaboration between them straightforward, and the nature of their role in local economies made it easy and advantageous for imports, currency, warehousing and marketing to be handled by others from the same group. Social gatherings were and still are common and the Yellow Pages in any Latin American city will list a series of social clubs that act as meeting places for people descended from immigrants from a particular area. This same phenomenon is, of course, characteristic of all migrant groups in the northern hemisphere as well as the south, but it does enable small but important immigrant groups to be identified in every Latin American country.

Movement between countries in the Americas

The arbitrary nature of many Latin American frontiers ensures that there is a great deal of movement of population across international borders, but the similarity of the cultural and ethnic composition of people either side of the border means that the movements are a reflection not of the nature of the people but rather of economic and political differences whose natures change from time to time. Where major cultural divisions may seem to exist such as between Latin and Anglo-America (Mexico and the USA) such differences are much less remarkable on the ground. Much of the southern border regions of the south-west of the USA were formerly part of Mexico and the population of the adjacent areas of the USA is Latin American in origin.

Two broad categories of movement can be identified. In the first place it is a basic principle of migration that people move in response to their perception of opportunities for increasing their well-being and that of their households. Thus it is to be expected that one current of migration is towards countries and regions within them where employment is available and wages are high. Thus a sizeable part of the labour force cutting sugar cane in northwest Argentina comes from adjacent parts of Bolivia where wages are low and employment opportunities more restricted. Many Bolivian seasonal migrants remained and subsequently moved to the larger and more varied source of employment represented by Buenos Aires. Similar migration occurred within the past 50 years from northern Colombia to Venezuela in association with petroleum extraction and the development of urban and industrial services associated with it, both in the early petroleum centres around Lake Maracaibo and later to the Caracas region. Such population movements are commonplace in Latin America.

A further category of movement involving smaller numbers of people is migration in search of new land. Many frontier areas in Latin America are settled by few people and where disparities exist in availability of land on either side of a frontier then people seek to acquire land where it is most easily available. Thus Brazilians acquire land in eastern Bolivia, Salvadoreans move to adjacent Honduras and native Indian Guatemalans have moved to settle the southern and western parts of

Belize. Such movements pose a threat to the recipient country since all countries fear territorial losses when citizens of a neighbouring country populate land close to their border: frontier conflict during the twentieth century has regularly taken place between Latin American nations.

A final category of international migration that is of contemporary importance is that to and from the USA and other industrialized countries. On the one hand this includes the migration of those with or wishing to acquire special skills through education, and on the other, workers prepared to sell their labour at a rate sufficiently below that prevailing to ensure rapid if illegal employment.

The migration of less-skilled Latin Americans, largely to the USA, even though some have gone to the Persian Gulf states, is a phenomenon that has most affected Mexico, although it has, in very specific contexts, also affected both Cuba and Haiti. The migration of Mexicans to the USA will be referred to later but it represents a major possibility for earning to tens of thousands of Mexicans at many different economic levels. The great economic differences on either side of the border mean that Mexicans can readily find casual and short-term employment in the USA which enables them to save money to send or take home and to finance an illegal crossing of the border and transportation in search of employment. This form of migration is a real alternative to internal migration within Mexico and in a variety of ways decreases pressure on Mexican resources, provides a large amount of foreign capital and benefits many people besides the migrants through the spending of migrant earnings within Mexico. It has negative effects also in the cultural dislocation felt by migrants forced to travel long distances and live in an alien and frequently hostile cultural en- vironment, even though many migrants live in migrant communities, often made up of people from their own part of Mexico. They acquire, at least in part, a lifestyle that is unsuited to rural Mexico to which many eventually return, and thereby creates dissatisfaction which is not easily alleviated except through further periods of migration.

Conclusion

A rapidly developing continent freed from one form of colonialism only 160 years ago has increased its population by natural demographic increase and by the willing acceptance of large numbers of people from other continents. The consequence of these population movements has been to create not only a very culturally diverse environment but also to stimulate the use of many different cultural traditions in the search for well-being. The predominance of the capitalist economic system and the division of society along lines that were determined largely by differences in access to resources has also facilitated the conformity of people of these many cultural origins to patterns of society that are well known in western Europe and Anglo-America.

The diversity of cross-border migratory movement during the present century has reflected the levels of economic growth of different nations as well as regions within nations. Since such movements are often well-documented and recorded by national census data, migration may be used as a mirror of the geography of social and economic change and is therefore worth more detailed examination at a national and regional level.

Migration in the rural world

Migration from the countryside

Migration from the countryside is commonplace throughout Latin America and has probably been so for several centuries. While some communities sought refuge from exploitation by minimizing their contacts with the outside, threatening Hispanic world, others could not and did not avoid such contact and sold goods and services, provided labour and produce as tribute in towns and cities and regularly traded and worked in ecological zones both different and far removed from their own. During the present century, population in many areas had finally recovered from the demographic catastrophe that followed the arrival of the Europeans in the fifteenth century with their strange and deadly diseases. For the past 40 years population has actually declined or is stagnant in many parts of rural Latin America. This situation is demonstrated for the five central and northern Andean republics in Fig. 7.1. What are the distinguishing features of areas of out-migration? Data from a wide range of sources indicate that the continued exodus is from rural areas with the following characteristics:

1. Few possibilities for producing a surplus of cash crops, or of earning wages or of starting a business.
2. Isolation from urban centres where goods might be sold or employment sought.
3. A lack of cultivable land – frequently this identifies areas with a high density of rural population.

Continued emigration is closely associated with a long tradition of out-migration and a powerful factor in individual decisions to migrate is knowledge of possible destinations and having friends and kin who can receive and help migrants in the initial post-arrival period. Migrants are both wealthy and poor, though seldom the poorest since some capital is needed to be able to leave; and migration occurs similarly from Indian and non-Indian areas in the long-settled lowland and highlands of Latin America.

The consequences of this migration for the sending communities are varied and certainly not all negative. Migrants return home with money or goods bought with their earnings. In cases where migrants are absent for long periods they send money home to support non-migrants. This money is spent locally, especially on improving housing, it stimulates commerce and may even lead to investment in agriculture. The land of

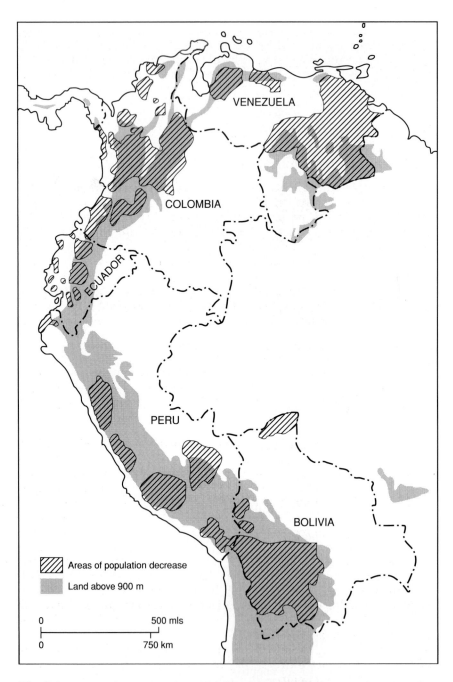

Fig. 7.1 Areas of out-migration in highland South America

Plate 7.1   Multi-storey house in Bolivian countryside 1994. Alternatives to migration are few. The opportunity for non-farm work may allow some to work away from home – in this case as pig dealers – but creates a quasi-urban life style with this three-storey house, in Batallas, Bolivia (Photo: D Preston)

migrants, if not farmed by remaining family members, is rented (or sold) to non-migrants, thus increasing their access to land resources. Returning migrants may come back with money and new ideas of ways to make a living, including some which may be relevant to farming. In many cases the communities from which migrants have gone had little land per household and have for a long time been dependent on a variety of non-farming sources of income, such as carpentry, hat-weaving and rope-making. There, the money and ideas that result from migration go into non-farming activities rather than into farming which is seen as the source of food rather than as a basic source of goods to sell. In a successful savings and loan co-operative in a village in highland Ecuador which had experienced much emigration, almost 80 per cent of the loans approved were for home improvements and commercial ventures.

Migrants go long distances and often cross borders. Migration to the USA is commonplace for rural Mexicans but also occurs for people from all Latin American countries. A study of migrants to the USA from the Mexican state of Jalisco (Cornelius 1979) showed that migrants on average sent home about $200 a month but were also able to save another $300 or more to bring home with them when they returned. A more recent study of US migration from southern Ecuador shows a similar level of remittances by migrants. The most frequent investment was in land for a house or for farming but the most successful often attempted to start a small business. A major effect of migration to the USA is to stimulate the migrants to wish to acquire a broader range of consumer goods and to participate in modern consumer society. However unusual the USA may be as a migration destination in Latin America as a whole, a variety of other studies of rural emigration towards major cities suggest very similar effects on migrants.

In communities with considerable out-migration it is possible to calculate the numbers and proportion of the population that is absent. In our own research in Ecuador perhaps 60 per cent of the generation of children born in 1950 now live elsewhere and studies in Argentina and Mexico have shown this to be a common proportion of people absent. Both women and men migrate but males predominate. They move to different types of destinations. Women are particularly likely to move to urban centres, and domestic employment is their most likely first occupation. Men migrate to urban centres too but are far more likely than women to move to other rural areas, particularly new land settlement zones in the lowlands.

Rural emigration has frequently been held to be detrimental to rural areas because it leads to the loss of the 'best' people and when it continues for several generations those that remain are the oldest, most conservative and least imaginative. There is abundant evidence to show that the characteristics of migrants are very similar to those of innovators which might suggest that little innovation might be expected in areas with a lot of emigration. However, one of the basic principles of migration is that every outflow generates a corresponding inflow. Since much migration is short-stay, particularly in Indian areas, then any community is likely to contain a number of returned migrants. Thus if the community does lose those with a broad world view, with the capacity to innovate and the willingness to take risks for the possibility of larger gain, it also gains with the return of migrants with money and experience gained when away. The community may therefore gain from their knowledge and experience. Furthermore, while migrants are frequently more educated than non-migrants, if their education has not fitted them for livelihoods that can be carried on in their home area, then their knowledge and ability is not crucial to their community. If a butcher or carpenter migrates, their village will only suffer from their going if there is no other butcher or carpenter left. Rural emigration does not thus necessarily cause a loss of people in the labour force who are essential to the maintenance of existing levels of living in the community. Furthermore, whether such migrants do constitute the 'best' people depends on value judgements that may be hard to verify.

The decline in population affects specific parts of rural areas in different ways: the absence of migrant males, for instance, affects specific tasks as varied as ploughing fallow land and occupying positions of authority in the community, for both of which male migrants are sometimes called home. The loss of younger people naturally affects reproduction rates although some migrants bear children while away and yet leave them in the village while they work elsewhere (particularly women employed in domestic service). The most profound consequence of migration is the exposure of many people and their kinsfolk to other ecological zones, to different ways of making a living and to other social and cultural values which they would not otherwise have known. A consequence of this increased circulation of people between towns, cities and rural areas is a need for cash with which to purchase what is newly felt to be essential. In the areas we have studied this did not lead to increased commercial farming although this has occurred in some places. It has led to a greater degree of social and political awareness and to a gradual proletarianization as people become involved in ways of production other than farming and handicrafts. Migration does not necessarily lead to the decline of rural areas. It leads to a redefinition of the space in which household livelihood is maintained. A rural-based household may become, as a result of migration, part of a complex multi-household livelihood strategy where co-operation between kin and friends in different locations enables economic and personal crises to be overcome, and security as well as profit are maximized.

The search for new land

A completely separate aspect of population mobility in relation to rural areas is that concerned with new land settlement. This is also the aspect which has traditionally received most attention from geographers for it relates to assessment of the agricultural potential of the physical environment. The process of new land settlement in Latin America has been important and well-documented for the past 150 years. In recent decades the settlement of vast areas of the Amazon basin associated with major road-building programmes embarked on by Brazilian governments has impressed many and also given rise to fears for the ecological future of this major region. The recent establishment of farms in other lowland humid tropical lands (from Guatemala to Bolivia) may also have led to major environmental changes, the consequences of which are still being experienced.

The search for new land has been associated with the changing needs of rural populations since prehistory. The movement of the Mayans from the limestone lowlands of Yucatán to the hills on the Central American cordillera, and of the Incas to settle and/or subjugate other highland and lowland areas are examples of new land settlement in the past. It is now widely recognized that many Andean peoples had access to land in different ecological zones which included both high-altitude grasslands and sub-tropical lowlands. It is the search for suitable lowland farmland that is the most important cause of most new land settlement in Latin America today. What concerns us here, in the context of population mobility, is the nature of the search for new lands, the origins of migrants, the nature of their destinations and the changes in the human geography that result from such migration.

The most important cause of the search for new land is that the would-be colonist feels that settlement in such an area will allow more freedom and opportunity both for the adults and for any children that they have or expect to have. For colonists from urban backgrounds this may be because of high rents and the fierce competition for employment or for clients in big cities; for rural people it may be because of land shortage, recurrent climatic crisis or low returns from domestic craft work. Others are escaping political strife (as in Peru during the rural violence of the 1980s and 1990s), or the police seeking them for serious crimes or parents for carrying off their daughter, even if not against her will. Some feel that a new social order will be created on the frontier where people will be judged by what they achieve rather than by their parentage.

Plate 7.2   Settlers in the Orinoco area of Venezuela (Photo: Tony Morrison, South American Pictures)

The geographical and other origins of colonists are varied. In some areas the majority of colonists are from rural areas, in others they are largely from small towns and even cities but in many cases they have at some time had experience of farming. Not all 'colonists' intend to become farmers, for some acquire land, farm it for a year and rent it to another family while they establish themselves as merchants or shop-keepers, only visiting the farm once a month or so and taking little part in its management. Many colonists come from areas where land is in short supply, typically from the highlands (in much of Central and Andean America) or from other areas where life is hard. Few colonists come without previous experience of the lowlands and many have migrated there as labourers on many occasions and thereby gained knowledge about the area and its possibilities. Many have kinsfolk already farming who can give guidance and, in some cases, colonists are moving from another newly settled area as the land there becomes less productive. A study by Connie Weil of one such area in the Chapare zone of Bolivia showed that four-fifths of the farmers had moved from another part of the same colonization zone although most had been born in the adjacent highlands (Weil 1983). Moreover, colonists often retain land in their home community in the highlands, partly as an insurance against failure but also because they intend to travel regularly between the two areas. In Weil's study colonists spent as much as 25 per cent of their time cultivating their other land whether in the highlands or in an older colonization zone. In Mexico and Guatemala highland families regularly migrate to work in the Gulf lowlands and some eventually obtain land there and cultivate crops in both zones.

New settlers come to the new lands in one of two ways. A minority come as settlers on directed colonization schemes. They are allocated a parcel of land, given advice on what to grow, loaned money to get over the initial period before the first harvest and generally given help to make a success of their new life. The majority are spontaneous colonists who have found their own land, either bought it from the previous owner or claimed it from the government's land office. They receive no help and there is no effective scrutiny of their past to attempt to determine whether or not they are suitable colonists. Large sums of money, often from international agencies, are spent on the former group – as much as $10,000 per family in some cases – to ensure success, and yet such schemes have a high turnover of settlers, more of them default on bank loans than the self-selected spontaneous colonists who tend to persist longer in their efforts to succeed. One of the reasons for this disparity in levels of success is that the government agencies directing colonization select poor sites for farmers and give little advice, which in any case may be inappropriate. Their criteria for selecting farmers are poorly chosen, and, perhaps most important of all, the directed colonists are tempted or forced to place too great a reliance on the managers of the project rather than allowed to follow their own judgement. Recent Mexican colonization experience has laid great emphasis on direction from above in the Gulf coast schemes, such that colonists feel themselves to be mere government employees. Other work in the Amazon Basin suggests that new colonists pay insufficient attention to the farming methods of the few indigenous or long-standing farmers who have, through a long process of trial and error, evolved a stable and productive farming system.

The most important factors in the success of colonists seem to be having some financial resources to support their enterprise. Those colonists with most resources or most land elsewhere were frequently those who enjoyed most success. They buy up the land of the poorer colonists and acquire sizeable holdings which, in the Santa Cruz area of Bolivia, facilitate the transfer to cattle rearing which is often the second stage of land use following forest clearance. There is no evidence to support the idea that the frontier really does offer a new social order, for the better-off become the new élite and employ the poor to work their land, often employing the former owner of land that they have bought, as sharecropper or tenant on the land they formerly owned.

Conclusions

Migration is a process that affects the majority of Latin Americans in both rural and urban areas. An important conclusion of migration research is that migration has no simple cause. It reflects major changes in society. One of the most important contemporary changes is the increasing demand by poor households for more consumer goods and access to a quality of life that they now realize is enjoyed by many people. Thus Bolivian rural people in the northern Altiplano or the valleys of Tarija near the Argentine border are buying gas cookers and televisions, and the more affluent are buying video-recorders, where 10 or 20 years previously cooking was exclusively by dried llama dung and wood, or kerosene among the more affluent. Such goods are paid for by wages earned when away from home and by increasing the amount of farm produce sold. These changes and the migration associated with them can most convincingly be accounted for by identifying the reasons behind the demand for more cash as well as the range of means by which such capital is acquired.

Migration and urban places

As an increasing proportion of Latin America's population is urban-based then migration to, from, within and between urban places will inevitably become a more important component of migration. Furthermore, since urban populations are growing rapidly in absolute terms and also experiencing profound socioeconomic change then the characteristics and consequences of migration involving urban locations are of particular importance. Since a widely held view of urban growth in Latin America is that it is largely a consequence of migrants swelling urban population, it is necessary to indicate the characteristics of migration to urban centres and to identify the specific role that migrants play in the dynamic urban geography of Latin American urban centres.

Characteristics of migration to towns and cities

Migration has a pronounced impact on the social, economic and political complexion of both the sending and receiving areas for migrants are not uniformly representative of all social groups. Likewise, if they differ markedly in a number of respects from the host population, friction may result and also the composition of the urban population will change. Although most of what has been written about urbanward migrants refers to cities, it is worth noting that migration occurs to almost any urban centre whether it be a small village, a regional capital or a metropolitan mega-city.

This can be illustrated by considering the actual migration patterns in one part of the Ecuadorian Andes. The parish centre of Quilanga is a large village with a population of 700 people but it is the social and economic centre of a rural area with a population of some 3,600 people. The parish centre attracts in-migrants from the outlying rural areas who seek what, in their words, is the more 'civilized' environment of the little town. In Quilanga, there are two single-sex primary schools with a teacher for each grade, a secondary school, a medical centre and a priest, as well as a number of shops selling a wide range of goods, several buses each day to the provincial capital of Loja and all the social intercourse possible in a small town at the end of the day and at weekends. This represents a better social and economic environment to people from the country districts.

Some of those who live in Quilanga town, however, find it depressing and intend to leave. They yearn for brighter lights, for the chance of jobs not related to farming. They want a secondary school that offers the complete secondary school curriculum instead of just the first part, they want to be able to buy a wider range of goods more cheaply, to have a home with electricity and running water and to be able to feel less isolated from the trends of life reported by the media. They envy kinsfolk in Loja with television and the young people want to be able to have the opportunity of going to the cinema or buying a drink in a bar where their behaviour will not automatically be noted and perhaps reported to their parents. So some of the small townsfolk go to Loja, a city of 66,000 people, where there are half a dozen buses a day to Guayaquil, the coastal metropolis, and to Quito, the capital and highland urban centre.

The people of Loja, for their part, are conscious of the small size of their city, that it is inaccessible except by a long, and relatively expensive road journey from the big cities and being a few hours from a recently opened road to the jungle is no compensation for the time and cost of getting to Guayaquil, a million city. They hear from relatives and on the radio and television of the range of jobs available in the big city, of the better quality of schooling, for all the best teachers and doctors want jobs there and they are well aware that politicians are more readily influenced by petitions from citizens in the big city than in the distant country areas. A teachers' strike in Loja would be relegated to the inside page of a national newspaper; one in Guayaquil would be worthy of a spot on the television news and on the front page of the main daily paper.

Thus there is a current of migration up and down the settlement hierarchy as individuals and groups seek to satisfy their aspirations and migration becomes an important dynamic component of population change in many categories of urban centres (Fig. 7.2).

It is axiomatic that migration flows contain a powerful element of momentum and many migrants are led by others from a similar area and with the same goals so that it is common that migrants from the

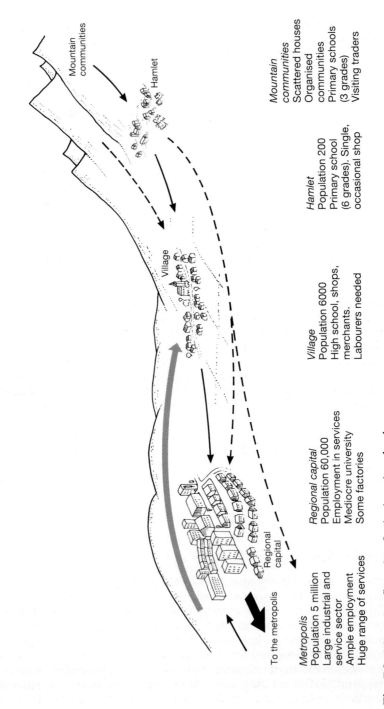

Mountain
communities

Hamlet

Village

Regional
capital

To the metropolis

*Mountain
communities*
Scattered houses
Organised
communities
Primary schools
(3 grades)
Visiting traders

Hamlet
Population 200
Primary school
(6 grades). Single,
occasional shop

Village
Population 6000
High school, shops,
merchants.
Labourers needed

Regional capital
Population 60,000
Employment in services
Mediocre university
Some factories

Metropolis
Population 5 million
Large industrial and
service sector
Ample employment
Huge range of services

Fig. 7.2 Migration flows in a Latin American landscape

same community or area will congregate in the same part of their urban destination. The extent to which migrants segregate themselves in particular geographical areas is much debated but in the period immediately after initial migration many of the migrants from any zone of origin are likely to be living in close proximity although as they acquire more knowledge about the town or city they move more independently. Where migrants from a particular area have tended to find similar sorts of employment, as, for example, flower-sellers in one Mexico City market have done, then residential propinquity may be maintained. Where there is a major cultural difference between migrants and urban population then migrants are more likely to stay together for reasons of solidarity until they can communicate on equal terms with the population as a whole. Some migrants also remain in close proximity in their urban destinations for commercial purposes.

The network of links that develops both between migrants and between migrants and their home areas is important for economic survival and for social satisfaction. These links are not necessarily based on migrants living close together but there are frequent occasions when migrants from the same area meet together to play football, to plan political action on behalf of the home community or to organize their contribution to major festivities either in the city (for a national celebration) or in the home community for a festival. These links enable migrants to weather periodic personal crises and the flows of goods are in both directions between newly arrived and long-standing migrants and between migrants and stay-at-homes.

Migrants are frequently different from the urban population as a whole. They are frequently younger than the average urban dweller, they often have a different range of skills and their level of education is lower than the urban population as a whole. Yet various studies show that migrants are relatively successful at finding jobs and making their way in the larger town. Although they may suffer discrimination if they belong to a particular ethnic group, they are not necessarily handicapped by reason of being recent migrants. Migrants too are not from any particular social stratum. Although many migrants from rural areas are relatively poor they often maintain resources at home. They seldom sell any land that they may own and better-off land-owners as well as poorer peasants migrate to urban centres.

Migration from and within urban areas

Migration from urban areas is not necessarily related to any deterioration of the quality of urban life. It does, however, increase at times of economic depression and decrease when jobs are plentiful and wages adequate. In addition there is an element of outward migration to establish businesses in smaller towns or villages and to use the information networks that have developed during years of urban living. Other urban out-migrants have tired of city life, of the high cost of living, especially housing, of crime and violence, of extensive atmospheric pollution and of the high pressure of city life. It is they who return to their village to set up a shop, to buy land or livestock and farm. It is also they who go to colonization areas. They thereby act as disseminators of urban values and lifestyles in far-off, non-urban areas.

Another increasingly common form of urban emigration is that of suburbanization and the creation of dormitory dwellings in the countryside on the periphery of the biggest Latin American cities in exactly the same way and for similar reasons as such urbanites seek rural homes in the industrialized nations. Caracas, Bogotá, Mexico City and Lima all have areas where the more affluent of the city choose to live, up to an hour or more's travel from the city centre. Such areas are not necessarily accessible by public transport but the presence of commuters gradually transforms the rural towns or villages as new residents seek and obtain better municipal services. Another trend, which affects the smaller cities, is the increasing number of people, both professionals such as teachers and manual workers, who travel from their village 20–30 kilometres from the city by bus each day to work in the city. Their cost of living is lower in the village than it would be in the city and they can bring their children up in a more stable and friendly social environment. Such suburbanization has been little studied in Latin American cities but to a visitor the spread of city folk into the rural areas is easily noted and is an important element in the changing urban geography of big cities.

A further feature of urban growth which is likewise associated with some of the largest Latin American cities is the rapid spread of the poorest households into the far periphery of the city. This inverts the customary social geography of Latin American cities where the poor settle the less desirable sites as close to the centre as possible – swamplands, hillsides, etc. – and it seems to have been the result of repressive policies of city and national governments that have placed severe restrictions on the growth of informal, semi-legal, low-income housing within the city periphery and thereby forced the poor to migrate from the city and make their homes in undeveloped areas beyond the city limits. Thus entering Caracas from the coast the first sign of the city is shantytowns clustered on steep hillsides several kilometres from the paved roads of the suburbs of the city. Several writers have documented similar phenomena in São Paulo where vacant land nearer the city centre that would, in a different political climate, be settled on by squatters is held by land speculators; the poor are forced out to the very margins of one of the most populous and sprawling cities in the hemisphere and thus are forced to travel for two to four hours daily to and from work from their homes on the very fringe of the city.

Further reading

Carlstein T (1982) *Time, Resources, Society and Ecology*, Studies in Human Geography No. 49. Department of Geography, University of Lund, Lund.

Cornelius W A (1979) 'Migration to the US. The view from rural communities', *Development Digest* **17**, 90–101. (Good short summary of extensive research on Mexican migration to the US which is still highly relevant.)

Crossley J C (1983) 'The River Plate Republics', in **Blakemore H** and **Smith C T** (eds.) *Latin America: Geographical Perspectives*, pp. 383–455. Methuen, London.

Faron L C (1967) 'A history of agricultural production and local organization in the Chancay valley, Peru in **Steward J H** (ed.) *Contemporary Change in Traditional Societies*, pp. 229–94. University of Illinois Press, Urbana. (An excellent, detailed account of the history of settlement and associated migration in a valley near Lima.)

Hägerstrand T (1957) 'Migration and area', in **Hannerberg D** (ed.) *Migration in Sweden* (Lund series in Geography, series B) **13**, 25–158.

Skeldon R (1990) *Population Mobility in Developing Countries*. Belhaven Press, London. (Good introduction to migration with a case study of Peru.)

Weil C (1983) 'Migration away from landholdings by Bolivian campesinos', *Geographical Review* **73**, 182–97.

CHAPTER 8

Caring for people: health care and education provision

Joseph L. Scarpaci and Ignacio Irarrázaval

Caring for Latin Americans in the 1990s reflects the interplay of myriad practices, institutions, and cultural legacies. The caring sector – limited in this chapter to health care services and education – plays a key role in enhancing well-being. The task, though, is daunting. In 1989 some 44 per cent of the Latin American and Caribbean population (183 million people) were classified as poor. Health care delivery is spotty and few countries guarantee comprehensive care for the entire population. An estimated 130 million Latin American poor have no access to health care. Getting into the modern biomedical system depends on a complex array of gatekeepers – financial and political – and the quality of care is highly variable. When governments allot funds for health resources, as much as 30 per cent of them are wasted. Patterns of health care break down along geographic, government, ethnic, religious and labour lines throughout most of the region.

The quality and scope of education also varies greatly. Public education, the main institution for schooling, is generally of low quality though coverage is quite high. Private primary and secondary schools facilitate occupational mobility but their high costs confine those services to the élite. From the early years of schooling, most students are locked into a system of education that shapes social reproduction in Latin America and the Caribbean.

Such a bleak overview of the caring of Latin Americans might speak poorly of the region's health care and education systems. However, the average Latin American lives longer and is more literate than her or his counterpart in Africa or Asia. What accounts for these patterns of health and education in Latin America? What challenges do the region's delivery systems face on the eve of the twenty-first century? To answer these questions, we begin with the colonial period and assess the historical roots of the region's well-being at the onset of Europeanization.

Colonial beginnings

The Church was entrusted with important humanitarian and cultural tasks in addition to the spiritual instruction of Spanish and Portuguese colonists as well as the indigenous population. It took responsibility for teaching and used institutions to interject theology as a main component in the curriculum. Charitable institutions, asylums, alms houses, and hospitals were administered by the many religious orders. Servants of the Church used education and caregiving activities to maintain Iberian culture and complement economic well-being.

Health care

Europeans encountered salutary conditions upon their discovery of the New World. Estimates of the native contact population range from a mere 60,000 to 8 million, consisting mainly of Aztecs, Mayas, Incas, Chibchas and Araucanians. Low-density Caribbean settlements, urban centres from Tenochtitlán to Cuzco, and other secondary centres of indigenous peoples, were described as healthy and knowledgeable in ways conquistadors could not fathom. Their knowledge of herbal remedies and surgical procedures was vast, but it was not able to provide any defence against the new bacteria and viruses introduced from Europe. The first outbreak of infectious disease, thought to be smallpox, was probably in Hispaniola in December of 1518. The conquistador was less interested in learning the ways of this New World medicine, and more concerned about protecting himself from infectious diseases. Like the infamous *cordon sanitaire* imposed by the Europeans in colonial Africa, the early Latin American settlement was quickly segregated along class, religions, racial and health lines. As a result, military engineers devoted more time to quarantining the infected European and Native American from the healthy population than trying to understand the cause of ill health in the Americas. Although there were many factors other than imported sickness that led to aboriginal demise, 'disease proved the most destructive agent of a fatal complex' (Lovell 1992: 426). Colonization signified, as Herring (1965: 153) noted, an existence for the Indian that 'was undoubtedly distasteful, always disruptive, frequently cruel, but it was not a shift from paradise to torment'.

Colonial schooling

Popular schooling for Spanish Americans relied heavily on the importation of books from the motherland. Andalucians, Galicians, Basques and Catalans had no choice but to read Castilian. Ironically, Castilian became a more universal language in Spanish America than in Spain. Class and gender determined who would read often and well, and the sons of prosperous *peninsulares* (Spaniards born in Spain) and *reinóis* (Portuguese born in Brazil), *mestizos* and Creoles benefited. As in

Spain, education was not available to all Europeans, and it was denied to Negroes and Indians. The provision of schooling remained haphazard and impoverished, as a result. By the end of the colonial period, most Spanish Americans were illiterate.

Religious orders and colonial education

The European cultural base of Spanish America was expanded by the Franciscans in Mexico. Departing from the élitist notion that only the offspring of the noble should be educated, the Franciscans believed educating the Indian would allow them to value European culture. Pedro de Gante, a Flemish Franciscan, founded the first school for Indians at Texcoco (Mexico City). Under his administration over 40 years, between 500 and 1,000 Indians annually learned decorative arts, carpentry and masonry. These 'trade schools', however, used artisan labour only for adorning Catholic churches. In Peru similar schools were established. Nonetheless, by the end of the sixteenth century this popular form of vocational-technical schooling had ended, and the colonies returned to teaching young men from privileged heritage. This shift represented a change from the early periods of conquest and its emphasis on prostelytization to more effective political control and economic production.

While Spain's contribution to public schooling in the colonial period was neglient at best, it introduced high-quality universities for the élite of the New World. The University of Alcalá de Henares and the University of Salamanca in Spain served as models for Spanish American universities. In medieval Europe, the University of Salamanca rivalled universities in Paris, Oxford and Bologna. Beginning in 1551 under the rule of the Hapsburg Charles V of Spain, more than a dozen minor institutions and ten major ones were established. Not surprisingly, the earliest settlements in Mexico gave rise to the most prestigious universities. The University of Mexico in the seventeenth century held 23 chairs (*cátedras*) in the fields of rhetoric, theology, surgery, anatomy, medicine, canon law, and Otomí and Aztec languages. In Lima, the University of San Marcos served as a major learning centre throughout the Andean region. Northern South America had the University of Antioquia (1822) and the Jesuit-founded Pontificia Universidad Javeriana in Colombia (founded 1622 by the Jesuits and re-established in 1931).

The colonial Latin American universities suffered from religious rivalry and disputes among ecclesiastical orders. Secular factions fought to interject their views in curricular design, admissions and examinations. In addition, civil authorities pressured faculties for admissions and favoured young men from élite families. In the remote areas of the Spanish empire, religious orders such as the Franciscans, Dominicans and Jesuits settled among small indigenous populations and colonized them. The Jesuits, banished from Spain in 1767, had their property seized and were sent to the New World. From Buenos Aires they set off for Paraguay where they arrived in 1588. They pacified and converted the Guaraní to Catholicism. 'Reductions' (*reducciones*) – about 100 of them – held up to 100,000 Guaraní who learned the ways and languages of the Old World, and also served as a military buffer for the Portuguese colonies to the east. Although the Indians ultimately deserted these missions, the Jesuit effort exemplified how education and colonization worked in tandem. It was not until the late eighteenth century that European influences began to weaken ecclesiastical control over education. This 'new enlightenment' set the stage for political emanicipation in the following decades. Thus university and popular education in colonial America served many purposes for Spain: economic, spiritual, military and ideological. Class, race and gender determined the quantity and quality of formal learning.

Nationhood and the caring sectors

The cultural and intellectual achievements fostered by the Spanish monarchs were neglected after the colonial struggle. The wars of independence in Spanish America in the early nineteenth century deterred material progress and improvements in health and education. Business languished, tax bases for funding church or state projects were debilitated, and many of the region's intelligentsia went to Europe. Scholars from Venezuela (Andrés Bello), Argentina (Domingo Faustino Sarmiento and Vicente Fidel López) and Colombia (Juan García del Río) sought refuge in Chile. Throughout Latin America and the Caribbean intellectual activity stagnated while the arts and education took a back seat to the political mastery sought by the fledgling republics.

It was not until the late nineteenth and early twentieth centuries that the great philosophers of Latin America would emerge after the intellectual stagnation wreaked by the wars. Government was in no position to respond. Like the United States and Europe during the nineteenth century, Latin American governments were expected to abstain from intervening in economic activities. Government's role was to provide a judicial process, a stable currency and military protection and to conduct foreign affairs. Leaders of the new republics had little interest in the caring sector. What little public monies remained after the war were pilfered by political corruption or directed to the maintenance of the armed forces. This created a new space for ethnic groups, organized labour and regional governments to participate in health care and education. The mutual-aid societies played an important role in providing such care.

Mutual-aid societies and caring in the Southern Cone: 1880–1945

Health care in Latin America during the eighteenth and nineteenth centuries differed little from patterns in North America and Europe. The wealthy could afford providers who, at the time, were deemed competent. These included barbers (bleeders), homeopaths, naturopaths and physicians. The advent of key discoveries in medicine, chemistry and anatomy boosted the standing of medical doctors throughout the Western world. This movement was enhanced by the rise of germ theory and positivism in the sciences. Gradually, these beliefs replaced assumptions that illness was caused by superstition (*el mal de ojo*) and supernatural ailments. Independent of the prevailing belief systems in the Americas and Europe, the seriously ill sought care in alms houses, poor houses, charitable institutions and hospitals.

The Southern Cone (southern Brazil, Uruguay, Argentina and Chile) was the first part of Latin America and the Caribbean to develop comprehensive caring systems in the form of mutual-aid societies. Because of the paucity of native American populations in this corner of southern South America, there were few competing indigenous belief systems about illness and well-being as were found in Andean- and Meso-America. Many European workers – mostly artisans and semi-skilled labourers – who were displaced by the industrial revolution and the mechanization of agriculture, migrated to the Southern Cone. Mutualism in the New World adapted readily to the capitalist mode of production which dominated Europe during the nineteenth century. Caring and charity were private endeavours limited to the wealthy. The working poor, on the other hand, relied on their resourcefulness. Bureaucratic relics of mercantilist period gave workers in the new manufacturing sectors an organizational structure to pool risks. Mutual-aid societies consolidated worker contributions for emergency funds for burials, disability, inheritances and, to a lesser extent, medical care. Although European immigration to the Southern Cone between 1880 and 1940 paralleled migration waves in the North Atlantic, immigrants in the former were rarely used as scapegoats for problems in the health care field. Thus, for example, the kind of scapegoating that was hurled on the Chinese immigrants in San Francisco was absent in the Southern Cone. While some European enclaves persisted in metropolitan areas in the Southern Cone (i.e. the Italian quarters of La Boca in Buenos Aires), these were temporary ethnic ghettoes that barely survived a generation or two.

The River Plate countries of the Southern Cone – Argentina and Uruguay – followed a pattern of mutual-aid society development similar to Europe. Argentine workers initially organized to protect themselves from the vicissitudes of death and sickness. These societies were organized along decidedly ethnic lines. Large Italian and Spanish hospitals founded last century still operate in Montevideo and Buenos Aires. Shipley (1977) found that at the peak of their existence in 1913, about a quarter of a million *porteños* (people from Buenos Aires) were contributing about 3 million pesos annually to these societies. The timing of this peak in mutual-aid societies in Buenos Aires comes some 40 years after the hiatus of similar groups in San Francisco, owing perhaps to the delay of industrialization in Latin America. There also emerged at this time resistance societies which, like mutual-aid societies, provided risk-pooling for their affiliates. However, these resistance societies would also call for strikes to gain concessions from capital. The aim of these societies in Argentina and Uruguay was to spread the word of the evils of capitalism and they would eclipse the rather narrow, welfare-related focus of mutual-aid societies during the 1920s and 1930s.

The Argentine labour movement, unlike those of Colombia and Venezuela, was well connected with the world economy between 1880 and 1920. It was for this reason that Argentina was also swept up in the labour-related issues emerging in Europe during the first decades of this century. The predominantly European immigrant labour group embraced the European developments in socialism, anarchism and syndicalism; the difference in the latter two being that syndicalism emphasized nonsectarian workplace organization to eliminate capitalist development. Connections between Argentine organizations and European movements were clear. The syndicalists were allied with similar-minded workers in France; anarchists had their counterparts in Spain; and the socialists were connected with social democrats in Germany and France. As the proportion of Argentina's GNP increased in the manufacturing sector, so the syndicalists and their unions gained strength. At the same time, mutual-aid societies began to dissolve and were absorbed by labour unions. This trend was accelerated when Juan Domingo Perón served as Secretary of Labour between 1944 and 1946. Perón's first presidency marked the clear concentration of caring functions – of the sort that had characterized mutual-aid societies at the turn of the century – in the hands of unions. The ethnic-based friendly societies were overshadowed by organized labour which, in turn, undermined efforts at increasing the role of the state in providing public medical care. Unlike the concerns of the over-production of practitioners in nineteenth-century England which threatened physicians by lowering wages, physicians in Argentina and Uruguay had to be more concerned with working with a powerful union or trade group which would, in turn, guarantee higher remuneration. This stands in contrast to the Chilean case, to which we turn.

The Chilean pattern of mutualism differed from its River Plate neighbours. The latter part of the nineteenth century was particularly cruel to the most destitute of Chile's poor. Chilean agricultural exports suffered greatly from the opening up of previously untapped expanses of arable land in Canada, Russia, Australia, the USA, India and Argentina. War with Peru and Bolivia (1879), the great depression of the 1870s, armed conflict with the Araucanian Native Americans in 1880, and civil war in 1891 wreaked havoc on the poor. Public relief for workers thrown out of work or for veterans and their dependants was meagre. This period was a time when Chile's two-tier class structure was most pronounced. On the one hand was a small mining, agricultural and commerce bourgeoisie which benefited greatly from Chile's insertion of primary commodities into the world market. On the other hand,

> hunger, unemployment, prostitution, crowding, unhealthy living conditions, exploitation, child abandonment, were the signs of a society sharply split. Infant mortality, the highest in the world, constituted the symbol of a social territory where the son of a proletariate had no historic legitimacy. Plagues

such as measles, cholera, alfombrilla – infectious diseases – typhoid, syphilis, tuberculosis – ruthlessly stalked the living (author's translation from Illanes 1989: 12).

Mutual-aid societies in Chile (*sociedades de socorros*) were formed among those who suffered from these social ills. They initiated pioneer efforts in public health and social welfare in their decisive role in enabling the people to avoid submitting to charity. Epidemics and chronic public health problems in Chile did much to uncover the poverty in Chile. Mutual-aid societies sought to insulate workers from these maladies. Legitimacy to their operations was gained by the support of the Democratic Party who pressed the national government to adopt laws that would provide health care and pension benefits for the working class. Their efforts, however, were defeated in the debates in Congress which included such statements as 'When the State interferes in favor of the weak and those without any inheritance … it does nothing more than repair the bad that was previously committed' (Illanes 1989: 27).

Despite the impressive growth of mutual-aid societies in the caring sector in the Southern Cone, they had lost most of their importance by the Second World War. In their place arose larger organizations based on industrial and manufacturing coverage (Argentina and Uruguay) and consolidated public programmes. The National Health Service in Chile was founded in 1952. Like many institutions imported from Europe, Latin American institutions and culture modified these care programmes in response to changing economic and political events.

Health care in Latin America

Health care delivery in Latin America is complex and geographically varied. An overview of these services in the twentieth century is given first, followed by a description of the main features of its availability and accessibility. The next sections examine health care expenditures, indicators of well-being commonly used to measure human welfare, and problems of health care among the urban poor.

Health care delivery in the twentieth century

Latin America stands out amongst the developing continents of Africa and Asia because of the early and active role of both the private and public health care delivery systems (Table 8.1). Perhaps because it shrugged off its colonial yoke much earlier than other corners of the developing world, a variety of public and private systems developed throughout the region. The quality and comprehensiveness of these systems, however, varies.

Health delivery systems manifest a society's distribution and organization of scarce resources. This distribution in the Third World frequently produces inequalities and a lack of social justice. As such, it

Table 8.1 Health service systems, by national economic level and health care policy among selected Latin American and Caribbean countries, circa. 1990

Economic level level*	Health care policy		
	Permissive (laissez-faire)	Co-operative (welfare)	Socialist (centrally planned)
Upper-middle income	Argentina Bermuda Uruguay Venezuela	Brazil Jamaica Mexico Trinidad and Tobago	
Lower-middle income	Bolivia Chile Paraguay	Colombia Costa Rica Ecuador Guatemala Peru	Cuba
Low income	Honduras	El Salvador Haiti[+] Nicaragua	

* Income groups used by the World Bank (1993)
+ More than 50 per cent of the funds for Haiti's health care system comes from international donors

Sources: Mercer Fraser (1991); Mesa-Lago (1992); Cristina Puentes, personal communication, Pan American Health Organization, March 1994

is easy to confuse individual health problems with societal problems. In Latin America there are wide gaps in the availability and accessibility of health care facilities. Availability refers to the presence or absence of such facilities whereas accessibility depends on the finances, location, race, class and gender of the health care consumer.

Availability and accessibility of health services

Cuba and Costa Rica today show the most comprehensive systems of health care delivery. Cuba finances health care through state funds and is the only nation in the region to provide comprehensive care. Comprehensive care refers to primary, secondary and tertiary care: primary care means diagnostic and therapeutic care provided by a doctor or nurse; secondary care denotes mainly general hospitalization, while tertiary care includes specialty hospitalization such as intensive-care units and other specialized services.

Costa Rica is typically described as a mixed system of health care that draws on private social-security institutions and public funds. During the 'lost decade' of the 1980s when personal incomes fell, national deficits rose and social services were scaled back, only nations with strong public health systems and nearly universal coverage for curative care were able to withstand the difficult period. Cuba, Costa Rica, Trinidad and Tobago were better able to keep their systems operating at customary levels. Curiously, the mode of health care financing throughout the region shows that no single type of payment system guarantees better access than others. However, both the Cuban and Costa Rican systems indicate that the state has a formidable role in delivering effective and universal health services.

The provision of health care is irregular throughout Latin America. The 1991 cholera epidemic in South America showed the inability of public health systems to cope with what once we considered to be an eradicated disease. The term 'public health' means services that are targeted to a large number people outside the 'clinical' setting. Typically, public health services deal equally with civil engineering (sewers, potable water, waste removal) and preventive care services (vaccinations, immunizations, pre-natal screening, monitoring during pregnancy). Cholera, however, first appeared in the coastal areas of northern Peru and then spread within the country as well as to Ecuador, Brazil and Colombia. By mid-1991 the disease had left 1,300 dead and nearly 3 per cent of the 22 million Peruvians had been infected. The point is that cholera – like many other infectious diseases endemic to under-development – thrives in weakened economies where public health systems react slowly. Neighbouring Chile, however, despite the privatization of large parts of its curative health care system, remained relatively immune from the cholera outbreak because of its long-standing public health services. The cholera outbreak of the 1990s underscores several key points about public health and economic development:

1. Gains in economic development across society can prevent public health disasters.
2. Investing in public health services and infrastructure is a low-cost strategy to deter calamity.
3. Improvements in public welfare in Latin America can be set back easily during periods of economic downturn.

Roemer (1985: 30–36) has conceptualized a useful framework for characterizing national health services. This framework consists of five components: resource development, programme organization, economic support, management and delivery of services. Resource development includes the human and physical resources. These consist of com-modities, facilities, knowledge and manpower. Programme organization determines whether the health services are entrepreneurial, voluntary non-profit or governmental. In most countries around the world, government health services almost always come under the realm of the ministry of health, and Latin America and the Caribbean are no exception. Economic support considers the various sources of financing: public, employer, out-of-pocket, charitable or collective non-profit. Management refers to the different types of control, including regulation, evaluation, administration and planning. These elements may be public or private. Putting these four components together gives rise to the last component: delivery of service. Health care delivery, as noted above, can be curative or preventive (public health), public or private, therapeutic or rehabilitative.

Table 8.1 combines these five health system components and presents an overview of the health systems in Latin America and the Caribbean. Most countries can best be characterized by the co-operative (welfare) model which draws on employer, employee and state-funded medical care programmes. Cuba is the only centrally planned health care system in the region and at the other extreme is the permissive (*laissez-faire*) model with greater private sector funding. It is noteworthy that except for Bolivia, Paraguay and Honduras, most countries in this group are more urban, literate and prosperous than the rest of Latin America. The three Southern Cone countries discussed above – Argentina, Chile and Uruguay – also fall in this category. Their experience reveals a shift from private and charitable mutual-aid societies to large institutions financed largely outside the public sector.

It should be noted that this is only a 1990 snapshot of health service systems according to economic level and health care policy. The situation is ever-changing. For example, the economic crisis in Cuba may place that nation's health care system outside the centrally planned economy if economic restructuring continues there. Likewise, Chile in 1980 would have been placed in the 'welfare' category but a rash of privatizations has reallocated its classification to the *laissez-faire* group. Roemer's framework enables us to look systematically at the inputs of caring for Latin Americans. Moreover, it underscores the diversity of models that exist.

Health care expenditures

The expenditures per head on health care were cyclical during the 1980s. Table 8.2 shows a decade of expenditures indexed at the year 1982 for the public sector in selected Latin American countries. The impact of the recession produced Draconian-like cutbacks as illustrated by Bolivia, Paraguay, Venezuela and Mexico, where per capita expenditures during 1983 and 1988 fell between 18 and 63 per cent. Conversely Argentina, Brazil, Ecuador and Uruguay managed to make steady increases in health care funding, while Costa Rica and Chile held intermediate positions. The weighted average at the bottom of Table 8.2 indicates that the region as a whole reduced expenditures in the mid-1980s and it was not until 1987 that expenditure levels were able to rebound to levels from the early 1980s. This dip in social spending characterized the 'lost decade' in Latin America.

Health care indicators

There are myriad indicators of health status and social well-being. Perhaps the most sensitive indicator of overall quality of life and health

Table 8.2 Latin America per capita weighted average health expenditures index (1982 = 100)

Country	1978	1979	1980	1981	1982	1983	1984	1985	1986	1987	1988
Argentina	181.6	158.2	173.6	146.6	100	113.1	137.3	132	184.2	196	141.5
Brazil	82.4	79.3	98.9	96.2	100	88.5	90.5	101.5	111.9	156.4	120.3
Chile	94.5	86.8	101.5	98.2	100	79.9	88.5	84.1	82.4	88.4	88.5
Costa Rica	111.7	120.7	135.8	114.3	100	87.9	98.7	86.3	85.3	94	104.6
Ecuador	74.5	73.3	94.3	107.1	100	85.9	87.1	89.9	93.6	128.2	105
Mexico	139.4	152	103.8	98.2	100	74.2	84.7	85.8	82.1	83.2	83.2
Paraguay	61.2	82.5	83.5	116.5	100	104.4	137.1	108.2	51.8	57.6	57.6
Peru	81.6	85.1	117.9	124.8	100	98.1	101.5	99	100.7	86.1	86.1
Uruguay	116.9	101.8	120.6	107.3	100	81.8	81.6	85.2	108.5	105.1	116.9
Weighted average	108.0	108.4	109.7	104.9	100	87.5	94.9	99.1	107.9	130.4	107.5

Source: CEPAL (1991)

services, however, is the infant mortality rate. This measures the ratio of infants who die during the first year of life to 1,000 live births in that same year. In general, there is a strong negative relationship between income levels and infant deaths. Thus, poorer countries have higher levels of infant mortality than wealthier ones. Table 8.3 uses World Bank data of low, lower-middle and upper-middle income countries in classifying infant mortality levels. The average 1990 infant mortality levels for these same groups are 71, 42 and 34, respectively, verifying this negative relationship between income and infant survival.

Table 8.3 Infant mortality rates in selected Latin American countries

Infant mortality rate per 1,000 live births	1960	1970	1980	1990
Haiti	182	141	110	94
Nicaragua	144	106	86	56
Honduras	145	110	83	49
Low income economies	**165**	**109**	**87**	**71**
Bolivia	167	153	126	83
Guatemala	92	100	66	60
Dominican Republic	120	90	65	54
Ecuador	140	100	78	47
Peru	163	108	83	53
El Salvador	136	103	72	42
Colombia	93	77	54	23
Paraguay	86	57	45	35
Jamaica	52	43	10	15
Costa Rica	74	62	18	14
Panama	68	47	33	21
Chile	119	78	27	17
Lower-middle income economies	**144**	**87**	**89**	**42**
Venezuela	85	53	39	34
Argentina	61	52	44	25
Uruguay	51	46	34	21
Brazil	118	95	73	58
Mexico	91	72	53	36
Upper-middle income economies	**101**	**72**	**58**	**34**
Latin America and Caribbean		**82**		**44**

Source: 1970 and 1990 figures: *World Development Report, 1993, Investing in Health: World Development Indicators*, Table 20, p. 292; 1960 and 1980 figures: *World Development Report, 1984* Table 23, p. 262

Health care among the urban poor

Despite the ways national health systems can be classified and the outcome measures available at the national level, there is a significant population that falls through the cracks and does not receive regular care. One of these groups is the urban poor. Many lack access to basic primary care. Whether sceptical of the quality of care in public settings, dissatisfied with long queues, or displeased by the quality of treatment by ancillary and medical staff and, many of the urban poor attempt to provide basic primary care to themselves and their neighbours. We call this care 'shantytown health care'.

Shantytown health care aims to take technical information provided by national and international NGOs and deliver locally. It draws on high levels of neighbourhood participation and focuses on relatively simple services. These services include monitoring low-birthweight babies during the early months, making sure pregnant women get pre-natal attention, and teaching the community about preventive health services such as immunization, contraceptive and reproductive services, and AIDS education. Although shantytown health care can never fully replace a comprehensive health care system, it does allow the urban poor to increase the quality of life in cities throughout Latin America and the Caribbean.

Health care delivery in the countryside is confined mainly to primary health care (provided mostly by non-medical personnel) and primary medical care (therapeutic and diagnostic care provided by nurses, midwives and physicians). Most small towns and villages lack bedded facilities for secondary care and other diagnostic services found at hospitals. In the absence of biomedical services, the countryside through-out the region hosts an array of herbal medical practices and beliefs (*mal del ojo, santería*). Referrals (*interconsultas*) in the modern medical system are usually made to medium-sized towns while the most sophisticated care (specialized tertiary facilities) is confined to the national capital or a few large cities. In this regard, the geographic hierarchy of villages, towns and cities parallels the range of health care services.

Educational levels in Latin America

We approach this section about education in Latin America with a brief revision of the recent trends in the educational indicators for the last decades. Perhaps the most remarkable feature is the systematic increase in levels of education that most countries have experienced. This note of optimism holds true for all levels of education and for most countries. However, inequalities remain between urban and rural populations and between men and women. In addition to these problems, an emerging issue in education is that of quality as measured by years of repeated schooling and low educational outcomes.

CEPAL (1991) documents an important increase in pre-school attendance during the 1980s. However, in the rural areas attendance was much lower than in urban areas. In the most successful cases, only one in five children in the relevant ages are attending nursery schools. There are also important inequities in relation to income groups. For example, urban Brazilian children belonging to the upper quartile of income are five times more likely to attend a pre-school than the poorest rural children.

During the 1980s the region also experienced an important increase in primary school attendance. This is also true even in areas where this has not been the case traditionally. In rural Brazil, for instance, the attendance rate increased from 69 to 87 per cent. However, school attendance is still closely linked to a household's economic condition. Students from poor families are often asked by their parents to work on the land or in household tasks. Thus, Brazilian children from the poorest households have attendance rates that are on average four times lower than those from the wealthiest households.

The overall evolution of the enrolment in primary education for the last three decades in Latin America is shown in Fig. 8.1. Countries are grouped according to the World Bank's income groups for 1993. The data show the highest increases in countries with low incomes: Nicaragua and Honduras. However, both countries still have enrolments below the Latin America average. Countries with lower-middle income also show improvements in primary enrolment above the Latin America average. This is true for Peru, Colombia, Ecuador and Chile. Finally, upper-middle income countries show increases in primary enrolment below the Latin America average. This is mostly explained by the slower trend of improvements seen in Brazil, which, because of its large population size, lowers the group average.

Women's enrolment has increased at the same level as the overall population in all income groups and in Latin America in general. Between 1970 and 1990, Latin American female primary school

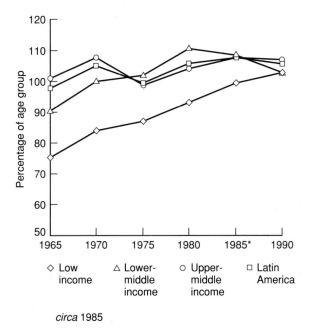

circa 1985

Fig. 8.1 Enrolment in primary education

enrolment rose by 12 percentage points to 106, two percentage points above the average enrolment of the high income economies.

CEPAL's (1991) report also computes an index of Household Educational Climate (HEC). This is the average schooling years of household members older than 15. According to the report, the HEC improved in Brazil, Colombia, Costa Rica, Uruguay and Venezuela: this provides a better capacity for socializing education, meaning that a household with higher HEC will be better able to capitalize on the educational opportunities available. However, the internal country differences remain high, as in the cases of Brazil and Venezuela.

Important gains in secondary education enrolment in the last three decades are shown in Fig. 8.2. The regional average rose from 19 to 49 per cent, more than three times the increase observed in primary education for the same period. This is because primary education was already widespread in 1965 (98 per cent), whereas secondary education was accessible only to the more privileged urban élites. Trends also show that the most important increases occurred in lower-middle income countries, specially in Chile, Peru, Ecuador and Colombia. In fact this group of countries show higher secondary coverage than the upper-middle income ones; this last group is again affected by the low coverage of Brazil and Venezuela.

The gender gap in secondary enrolment has narrowed in Latin America. Female enrolment increased 10 percentage points more than the total population between 1970 and 1990, reaching 57 per cent. This

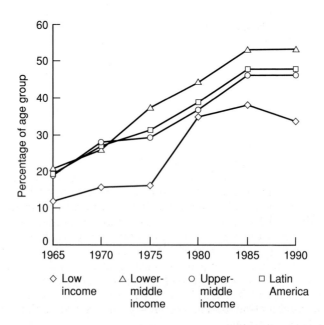

Fig. 8.2 Enrolment in secondary education

derived from an important expansion of secondary registrations, but it can also be combined with the effect of a larger proportion of young men dropping out of school to work. In fact, there is evidence that this situation occurred particularly in the mid 1980s during the economic crisis of the external debt.

Socioeconomic inequalities in education are noted in Table 8.4. The data contain information about the average schooling for youngsters, for the total and for the first and fourth quartiles. Although the average years of schooling are higher in metropolitan and urban areas, the inter-quartile differences are also the highest. At the same time, rural areas show less schooling but smaller inter-quartile inequalities. Thus, the geographic accessibility of primary and secondary schools is diminished even for the high-income rural households in Latin America.

Table 8.4 Average schooling years of young population (15–24) (selected Latin American countries)

Country	Area	Year	Total	Quartiles Q1	Quartiles Q4
Brazil	Rio–São Paulo	1987	6.5	4.7	9.1
	Urban	1987	6.6	4.9	8.8
	Rural	1987	4.2	3.3	5.5
Colombia	Bogotá	1986	8.6	7.4	10.0
	Urban	1986	7.8	6.8	9.1
Costa Rica	San José	1988	8.9	7.6	10.5
	Urban	1988	8.7	7.4	10.1
	Rural	1988	6.7	6.2	7.4
Venezuela	Metropolitan area	1986	8.9	7.9	10.8
	Urban	1986	8.0	7.5	9.2
	Rural	1986	5.7	5.2	6.6

Source: CEPAL (1991).

Literacy trends

In this section we briefly review Latin American trends in literacy because literacy remains a standard indicator of educational standards. In Latin America, there has been less variation in literacy levels in recent years. Table 8.5 shows that literacy throughout the region has increased by nearly 20 points, with a regional literacy rate of 85 per cent. Most upper-middle and lower-middle income countries show literacy levels above the region's average. This figure, though, is lowered by Brazil's poor performance. Latin America's poorest countries – Nicaragua, Bolivia and Honduras – have made remarkable gains in literacy levels which have narrowed the regional disparity. Overall, these improvements are not so much the outcome of special literacy programmes as they are a combination of the region's youthful population and high levels of school enrolment.

Table 8.5 Literacy trends in selected Latin American countries (1960–90)

Country	1960	1970	1980	1990
Nicaragua	38	57	90	91
Honduras	45	61	60	73
Low income	**42**	**59**	**73**	**81**
Bolivia	39	40	63	77
Ecuador	68	69	81	86
Peru	61	72	80	85
Colombia	63	74	81	87
Paraguay	75	81	84	90
Costa Rica	84	89	90	93
Chile	84	90	91	93
Lower-middle income	**65**	**73**	**81**	**87**
Venezuela	63	82	82	88
Argentina	91	93	93	95
Uruguay	90	91	94	96
Brazil	61	64	76	81
Mexico	65	76	83	87
Upper-middle income	**66**	**72**	**81**	**85**
Latin America	**66**	**72**	**81**	**85**

Source: World Bank, *World Development Reports 1983–1993*

In sum, the region has improved its levels of education during the past three decades. Although these improvements have been widespread across the whole population, the more vulnerable groups have been lagging behind. Despite a few programmes to combat adult illiteracy, there are not very many examples of major structural improvements in the Latin American educational system.

Educational expenditures in Latin America

Before introducing the trends in educational expenditures in Latin America, it is important to note the evolution of the public sector in the recent past. Latin America has been shaken by two severe economic recessions in the past 20 years. One was the oil crisis in the mid 1970s and the other was the external debt crisis that hit in the 1980s. As a result of these economic conditions most countries adjusted their economies by shrinking public expenditures. Real social expenditures per head fell during the 1980s in most Latin America countries. However, this decline is not necessarily the result of new government priorities. Rather, social expenditures stem from a falling GDP that was not accompanied by reallocated spending.

The evolution of Latin American educational expenditures is shown in Table 8.6. The trends observed are very similar to those of health, but with a more constrained recovery at the end of the period. The table depicts three distinct periods: 1978–1982 which corresponds to a pre-crisis and recovery from the earlier oil crisis, the 1983–85 debt crisis, and the 1986–88 post-crisis. A few countries – Argentina, Brazil and Costa Rica – show a considerable recovery in educational expenditures in the post-crisis period. Because of its large size, Brazil influences the overall average. The rest of the countries surveyed – Chile, Ecuador, Mexico, Paraguay, Peru and Uruguay – had not been able to recuperate from the pre-crisis period.

Educational expenditures were somewhat protected during the 'lost decade'. This finding is supported by coefficients of vulnerability for educational expenditures produced from two studies. These coefficients measure the changes in educational expenditures in relation to the changes in the overall government expenditures, they are estimated in the context of shrinking public expenditures. The results of both studies are very similar, obtaining coefficients of 0.56–0.58. This suggests that education expenditures were to some extent insulated during the periods considered. In practical terms this means that education expenditure has decreased 40 per cent less than total government expenditures. Educational expenditures, however, were less protected than health.

A striking paradox arises from the spending patterns above. On the one hand, the data portray an optimistic picture of the educational improvements that Latin America has achieved during that last decades. On the other hand, we also examined the significant decreases in educational expenditures. An analyst from the World Bank (Grosh 1990) suggests four explanations for this:

1. Most changes in educational expenditures affect teachers' salaries but not always the quantity of inputs.
2. It is also possible to find lags behind the changes in inputs.
3. The educational system can de-capitalize over time, but some inputs are irreversible. For example, once immunized, a child is no longer receptive to the disease.
4. It is also possible to expect that during a critical period of government sector shrinkage, the private sector can respond to these trends. This private component includes the household, with higher contributions in money or time, and also contributions by NGOs. While these four points may be cited as apologies for the abdication of government's role, it highlights the conundrum brought on by maintaining external debt and economic downturns.

Immediate increases in social expenditures are unlikely for many countries that are just starting in the 1990s to recover from the recession. Therefore, improvements in service delivery will have to come from increasing equity and efficiency of the available resources.

Emerging issues in Latin American education

In this section we describe some of the most important issues in the provision of education in selected countries. The topics selected result

Table 8.6 Educational expenditures per capita: selected Latin American countries (1982 = 100)

Country	1978	1979	1980	1981	1982	1983	1984	1985	1986	1987	1988
Argentina	133.5	125.0	144.9	144.6	100.0	101.5	117.1	100.9	94.9	102.7	104.0
Brazil	86.9	89.7	66.4	76.5	100.0	71.3	62.2	76.8	101.9	130.9	140.0
Chile	85.9	87.7	89.8	100.6	100.0	86.3	88.9	88.1	86.9	83.3	76.7
Costa Rica	161.3	171.5	169.8	131.7	100.0	113.6	108.2	104.4	113.6	164.8	127.4
Ecuador	70.8	71.7	120.5	111.4	100.0	88.7	85.3	88.3	95.4	85.6	75.1
Mexico	70.0	73.1	77.9	95.6	100.0	67.4	69.8	72.5	60.3	61.3	
Paraguay	92.8	84.5	89.8	91.8	100.0	94.2	78.0	68.2	65.7	70.2	
Peru	77.1	60.2	103.5	114.2	100.0	87.5	89.3	85.3	113.1	79.5	
Uruguay	79.6	87.1	93.1	93.5	100.0	67.1	55.7	57.7	69.3	81.5	88.4
Weighted average	86.8	87.1	84.4	94.3	100.0	76.3	74.0	79.5	89.2	101.7	82.9

Source: CEPAL (1992a)

from the availability of information, rather than a precise selection of case studies. Even though they are presented separately below for the sake of a better understanding, most of the issues we describe are interrelated. We build upon the previous sections by examining the increase in enrolment and the decrease of state finance.

Redistributive impact of educational expenditures: regressive

The redistributive impact of educational expenditure addresses the question: who benefits from public education subsidy? It is usually assumed that the objective of state expenditures should be to help those whose ability to pay is low and who might face difficulties in attending school. In other words, the allocation of social expenditures should be progressive – they should improve the conditions of the poor.

In the case of Venezuela, Marquez *et al.* (1993) point out that pre-school, primary and middle education have a progressive impact on the socioeconomic conditions of the population. However, higher education expenditures, which take an important share of the educational budget, have a regressive impact. Students from the upper quintile income group receive about 40 per cent of educational spending, while those from the lowest 40 per cent receive less than 30 per cent. Similar trends exist in Peru (Briceño *et al.* 1992), where the poorest 20 per cent of the population received 17 per cent of educational expenditures. Conversely, the top 20 per cent received 15 per cent. In other words, Peruvian educational expenditures are progressive only from the lowest 40 per cent of the income distribution upwards.

The redistributive impact of educational expenditures has encouraged an important discussion about the most equitable allocation of public expenditures in the sector. There is a consensus that education is a key factor in upward social and economic mobility; therefore, the state has to assess which socioeconomic groups require the most state assistance.

Regional inequities

A related issue to the redistributive impact is the regional impact of educational expenditures. In some countries the overall impact of public spending can be progressive but it can hide important spatial inequities. Larranaga (1992) mentions that, in rural areas of Bolivia, there is only an adequate number of primary schools. Most children in those areas will have access to only five years of education, while enrolment in urban zones is more than double that of the rural zones. A spatial inequity results because the Bolivian population is almost equally divided between rural and urban areas.

The Peruvian Departments of Apurimac, Huancavelica and Ayacucho have rates of illiteracy above 40 per cent yet receive only 6 per cent of educational public expenditures. Additionally, although the lack of adequate infrastructure is a common feature in this country in where some 64 per cent of the educational establishments lack water, electricity or sewerage, in the rural sector this figure is above 80 per cent. The traditional discrimination in the allocation of public expenditures to poor rural regions has increased as a result of the guerrilla actions of Shining Path (*Sendero Luminoso*), which sometimes make state intervention impossible.

Expanding the educational system in Latin America and the Caribbean has benefited the cities more than the countryside. Enhancing the educational system usually starts in the capital and then spreads out to a country's secondary cities. The extension of these improvements to the countryside is usually limited or non-existent. Most Latin American countries have favoured education in the cities which, in turn, explains the high correlation between urbanization and literacy as well as urbanization and high levels of formal schooling.

Management weaknesses and centralization

The decreases in state expenditures in the social sectors have increased the discussion in policy and academic circles about the redefinition of the role of the state. The Costa Rican state has a long tradition of social service provision in one of the most stable democracies of Latin America. Nonetheless, according to Trejos (1993), since the Ministry of Education has the double role of provider and policy definer, it has resulted in an over-centralization of responsibilities. These roles have created an excessively rigid administrative procedure and an inability to define sectorial priorities and allocate resources correspondingly. As a result of this situation, Trejos argues for a more decentralized educational system that would allow greater participation and decision making at the local level.

A similar problem appears in the Dominican Republic. Santana and Rathe (1993) claim that the educational system lacks a normative administrative model. This means there is no match between the legal design of the educational system and the public funding which should be behind the education system. This mismatch between law and service delivery is further exacerbated by the uneven conditions in which teachers must work. For example, there are no standard levels of salaries, specified vacation periods, or criteria for promotion and advancement.

Poor quality of education

Poor quality in Latin America's public education system results from several factors. First, it is a system that is still in the early stages of development and, in most cases, has experimented with a variety of modes of provision. Second, in historical terms, the average per capita expenditures are less than a quarter of that spent in industrialized countries. This results not only in poorly paid teachers but also in few teaching materials. Third, the financial constraints of the 'lost decade' have produced a variety of procedures to adapt the system to the adverse conditions. Fourth, as mentioned above, the significant expansion in enrolment over the previous decades could have affected the quality of the system since more students from diverse backgrounds make teaching more difficult.

In the Costa Rican case several procedures helped to maintain the same enrolment levels in a context of financial constraint. More informally trained teachers have been added and the length of the school year has been decreased. Also, school building maintenance has suffered and in 1991 only 55 per cent of the students had the required textbooks.

Larranaga (1992) explains that the low wages for Bolivian teachers translates into poor-quality outcomes. In 1987, the real wages of teachers was 73 per cent of the 1980 level. However, their total income is often improved through taking a second job, since the average work day is less than four hours per day. Many teachers take a second job in private schools which boosts their income. The pattern of working in the public system for a small base salary in one part of the day and then the private system in the other is also common among physicians throughout Latin America and the Caribbean. What is needed in Bolivia and elsewhere, though, are improved systems to define the quality of education. Most countries approach the problem through the use of proxy variables such as repetition, teacher–student ratios and others. These variables are outcome indicators but do not explain the real causes of the problem of low-quality education.

Plate 8.1 Whatever the inadequacy of rural schools, teachers and pupils are proud of their achievements. Teachers and pupils in Batallas (Bolivia) smartly dressed in a formal procession, 1989 (Photo: D Preston)

Privatization and alternative modes of finance

Although privatization has become a very fashionable word in most Latin American countries after the economic adjustment of the 1980s, there are not very many examples of the privatization of the educational services. Instead, however, privatization often happens by default. In Peru, for instance, there is some evidence that the share of private sector expenditures in education has increased as a result of the decreasing expenses by government. In 1980 Chile introduced vouchers in its educational system. Until then, the Ministry of Education was the main provider of education in Chile and more than 80 per cent of students were enrolled in public schools.

The first element of these changes consisted in the transference of the control of the public pre-school, primary and secondary schools to the municipalities. This transfer was aimed at improving administration, increasing enrolment and upgrading the quality of education. The second element changed the traditional system of fixed-budget finance of schools to a per-student subsidy system aimed at promoting competition among the municipalities. It is also designed so private (subsidized) schools can attract and retain students since enrolment is the basis for obtaining financing from the central government. Under this system, municipalized as well as privately subsidized schools receive a per-student payment. The payment, the same for both types of school, provided a great incentive for the private (subsidized) schools, whose expenditures were, according to the Ministry of Education, 70 per cent of the average amount spent by the ministry. The payments varied according to the level of education, grade within a given level, type of education (elementary or secondary) and type of school (day or evening). In addition, the amounts differed by region to provide a hardship allowance for teachers working in remote areas. The individual per-student payments were initially related to a monetary unit indexed with inflation (*Unidad Tributaria Mensual*). In 1987, a new indexing unit was introduced (*Unidad de Subvención de Educación*) that would increase with the wages and salaries of the public sector.

Non-formal education in Latin America

The roots of the Latin American non-formal education movement derive from the works of Paulo Freire. His seminal works in 1965, *La Educación como Práctica de la Libertad* and *Pedagogy of the Oppressed*, have been widely read throughout Latin America. Non-formal education has been defined as 'any organized, systematic, educational activity carried on outside the framework of the formal system to provide selected types of learning to particular subgroups in the population, adults, as well as children' (LaBelle 1986: 2). Although non-formal education originated in Brazil in the 1960s, it spread to the rest of South America during the 1970s, following other social movements. Non-formal education has always been promoted by grassroots organizations that teach the poor to take control of their own lives. In the 1980s, popular education reached Central America. The case of the Sandinista government organizing a massive literacy campaign with more than 10,000 volunteers illustrates the potential reach of non-formal education.

The main goal of non-formal education is to enhance the power, status and quality of life of the poor. This is achieved through a set of guided experiences that comprise the participation in community-based programmes, the cultivation of personal attitudes, and the diffusion of information and skills (see Box 8.1). This is how non-formal education can be self-initiated or externally directed. Non-formal education is more than 'reading, writing, and arithmetic'; it is education for critical consequences. Seen from this perspective and in line with Freire's method, students participating in literacy programmes learn to read and write through discussions of the basic problems they experience.

Box 8.1 Features of non-formal education in Latin America

1. It demands a high level of participation
2. The concrete experiences of the learner serve as a starting point: everyone teaches; everyone learns.
3. It focuses on group rather than individual solutions to problems. It is a collective effort.
4. It stresses the creation of new knowledge rather than just transmitting existing knowledge.
5. It leads to action for change.

An example of popular education is the 'Popular Councils' in the municipality of São Paulo, Brazil. These councils were created to democratize local decisions. They try to channel the opinion of neighbourhood associations, trade unions, mother groups and other local organizations to bring about change. In Chile, human rights workshops conducted in the shanty towns were both efforts to inform Chileans about the status of the missing and disappeared, and to understand the workings of the dictatorship. The Brazilian and Chilean examples illustrate how non-formal education strives to instruct yet also enhance the disenfranchised by pointing out social and political issues.

Non-formal education has faced at least two setbacks since its beginnings in the 1960s. First, military rule in Latin America and the Caribbean has not been kind to non-formal education because it views many of its programmes as extensions of left-wing political parties. Although considerable momentum of non-formal education was lost during military rule in Brazil (1964–85), Argentina (1976–83), Uruguay (1973–85) and Chile (1973–90), it survives in various forms today. Second, the movement is facing a redefinition given the economic crisis of the 1980s. According to the conclusions of a popular education conference held in La Paz in 1990, popular education built a discourse related to the structures of domination in society. In doing so, though, it left aside the analysis of problems in daily life. They also concluded that there is a lack of systematic research that assesses the outcomes of popular education. In the light of these findings, the conference concluded that popular education must redefine its role, especially in light of shrinking government support for public education. The popular education movement realizes it cannot replace the state but it can help to articulate the kinds of demands placed on the state.

Conclusions

The caring sector in Latin America and the Caribbean – health care services and education – has responded to changing institutional, economic and political demands over the past 500 years. Health care has evolved from a charitable programme, to friendly societies in some parts, to highly institutionalized channels. Today, most health care delivery systems are either a welfare or *laissez-faire* type, with the increasing trend of including more private services. Cuba is the only centrally planned model. Latin America and the Caribbean remain the healthiest and most literate of the developing realms, surpassing regional averages of most indicators of well-being in Africa and Asia. There is a strong and positive relationship between the levels of urbanization and literacy and a decidedly negative relationship between income and infant mortality rates.

Despite poverty and pressures from external debt, Latin America has made important strides in improving access to education over the past 30 years. Higher enrolment ratios for primary and secondary students as well as literacy statistics support this claim. These programmatic changes have been progressive because the poorer countries have experienced above-average improvements. However, the narrowing of these performance measures in Latin America and the Caribbean has not eliminated regional inequities. Access to quality educational services is impeded along urban–rural, gender, racial, ethnic and socioeconomic lines. This is an important caveat in analysing national-level data because the above cleavages are often covered up as a statistical artefact. Clearly, though, Latin America's educational institutions have reached a significant level of survival. Popular education, also known as non-formal education, held great promise in the 1960s when it first spread from Brazil, but a changing regional economy, new interpretations about the role of the state, and periods of military rule have weakened the popular education movement. The challenges next century will be focused on matters of quality and designing curricula that address quality of life issues and meaningful participation in the community and workplace.

Further reading

Adeo C and **Larranaga O** (1993) *Sistemas de entrega de los servicios sociales: La experiencia Chilena*, Serie Documentos de Trabajo No. 152. Banco Interamericano de Desarrollo, Washington, DC.

Belmartino S and **Bloch C** (1989) 'Estado, clases sociales y salud', *Social Science and Medicine* **28**, 497–514.

Berquist C (1986) *Labor in Latin America: Comparative Essays on Chile, Argentina, Venezuela and Colombia.* Stanford University Press, Stanford. (Historically informed overview of labour–state relations in selected countries.)

Briceño A, Pasco-Font A, Escobal J and **Rodríguez J** (1992) *Gestión pública y distribución de ingresos: tres estudios de caso para la economía peruana*, Serie Documentos de Trabajo No. 115. Banco Interamericano de Desarrollo, Washington, DC.

Castañeda T (1992) *Combating Poverty: Innovative Social Reforms in Chile During the 1980s.* ICS, San Francisco.

CEPAL (Economic Commission for Latin American and the Caribbean) (1990) *Magnitud de la pobreza en América Latina en los años ochenta.* CEPAL, Santiago.

CEPAL (1991) *La equidad en el panorama social de América Latina durante los años ochenta.* CEPAL, Santiago.

CEPAL (1992) *Gasto Social y Equidad en América Latina.* CEPAL, Santiago.

Cook N D and **Lovell W G** (1992) (eds.) *'Secret Judgments of God': Old World Disease in Colonial Spanish America.* University of Oklahoma Press, Norman. (Comprehensive study of epidemics during the conquest.)

Finch M (1981) *The Political Economy of Uruguay since 1871.* St Martin's, New York.

Grosh M (1990) *Social Spending in Latin America. The Story of the 1980s.* World Bank Discussion Papers No.106. World Bank, Washington, DC.

Henige D (1978) 'On the contact population of Hispaniola: history as higher mathematics', *Hispanic American Historical Review* **58**, 217–37.

Herbert H (1965) *A History of Latin America.* Knopf, New York. (A classic and very readable history of Latin American government and politics.)

Illanes M A (1989) *Historia del Movimiento Social y de la Salud Pública en Chile: 1885–1920.* Colectivo de Atención Primaria, Santiago.

Knox P and **Bohland J** (1991) 'Social ecology, benevolent societies, and medical care delivery'. Paper presented at the meetings of the Association of American Geographers (AAG), Miami, 5 April.

La Belle T H (1989) *Non-formal Education in Latin America and the Caribbean.* Praeger, New York. (Assesses educational strategies outside the classroom throughout the region.)

Larranaga O (1992) *Macroeconomics, Income Distribution and Social Services: Bolivia during the 80s*, Cuadernos Serie Investigación No. 1-50. ILADES, Georgetown University, Santiago.

Larranaga O (1992) *Macroeconomics, Income Distribution and Social Services: Peru during the 80s*, Cuadernos Serie Investigación No. 1-49. ILADES, Georgetown University, Santiago.

Loudon I (1986) *Medical Care and the General Practitioner, 1750–1850.* Clarendon Press, Oxford.

Lovell W G (1992) 'Heavy shadows and black night: disease and depopulation in colonial Spanish America', *Annals of the Association of American Geographers* **82**, 426–43.

Marquez G (1993) 'Fiscal policy and income distribution in Venezuela', in **Hausmann R** and **Rigobon R** (eds.) *Government Spending and Income Distribution in Latin America*. IESA, Inter American Development Bank, Washington, DC. (Empirical overview of state expenditures in Latin America.)

Mera J (1988) *Política de Salud en la Argentina*. Hachette, Buenos Aires.

Mercer W M (1991) *International Benefits Guidelines 1991*. William M. Mercer Companies, London.

Mesa-Lago C (1992) *Health Care for the Poor in Latin America and the Caribbean*. Pan American Health Organization, Inter-American Foundation, Washington, DC.

Neri A (1983) *Salud y Política Social*, Hachette, Buenos Aires.

Rivas P (1988) 'Comparative aspects of health care delivery in Costa Rica and Cuba'. PhD dissertation, Cornell University.

Roemer M (1985) *National Strategies for Health Care Organization*. Health Administration Press, Ann Arbor, Michigan. (Descriptive and comparative review of how national health care delivery systems are set up.)

Romero H (1977) 'Hitos fundamentales de la medicina social en Chile', in **Jiménez de la Jara J** (ed.), *Medicina Social en Chile*, pp. 11–86. Editorial Aconcagua, Santiago.

Santana I and **Rathe M** (1993) *Sistemas de entrega de los servicios sociales en la República Dominicana: Una agenda para la reforma*, Serie Documentos de Trabajo No. 150. Banco Interamericano de Desarrollo, Washington, DC.

Scarpaci J L (1991) 'Primary health care decentralization in the Southern Cone: Shantytown health care as urban social movement', *Annals of the Association of American Geographers* **81**, 103–26. (Comparative look at poor people's health care in Montevideo, Buenos Aires and Santiago.)

Shipley R E (1977) 'On the outside looking in: a social history of the Porteño during the golden age of Argentine Development, 1914–1930'. PhD dissertation, Rutgers University, New Brunswick, New Jersey.

Torres R M (1988) *Educación popular. Un Encuentro con Paulo Freire*, Bibliotecas Universitarias. Centro Editor de América Latina, Buenos Aires.

Trejos J D (1993) *Sistemas de entrega de los servicios sociales: Una agenda para la reforma en Costa Rica*, Serie Documentos de Trabajo No. 153. Banco Interamericano de Desarrollo, Washington, DC.

Ugalde A (1992) 'The delivery of primary health care in Latin America during times of crisis: issues and policies', in **Weil C** and **Scarpaci J** (eds.) *Health and Health Care in Latin America during the Lost Decade: Insights for the 1990s*, Minnesota Latin American Series No. 3, Iowa International Papers Nos. 5–8, pp. 85–122. Prisma Institute, Minneapolis.

Van Dam A, Martinic S and **Peter G** (1990) *Educación popular en América Latina. Crítica y perspectivas*. CIDE, Santiago.

Waisman C (1987) *Reversal of Development in Argentina: Postwar Counterrevolutionary Policies and their Structural Consequences.* Princeton University Press, Princeton, NJ.

Weil C and **Scarpaci J L** (eds.) (1992) *Health and Health Care in Latin America During the Lost Decade: Insights for the 1990s.* Minnesota Latin American Series No. 3, Iowa International Papers Nos. 5–8. Prisma Institute, Minneapolis. (Reviews the successful experiences of health care systems during the economic downturns of the 1980s, with appropriate policy recommendations.)

Worcester D E and **Schaeffer W G** (1971) *The Growth and Culture of Latin America*, 2nd edn. Oxford University Press, London.

World Bank (1979–1993) *World Development Report.* Oxford University Press, New York.

CHAPTER 9

Industrialization and urbanization

Robert N. Gwynne

Latin American economies have been transformed by industrial expansion during the twentieth century, and particularly during the latter half of the century. This chapter will be broadly structured in terms of a structuralist or political economy approach to regional industrial development in Latin America. At this general level, the premise is that global economic processes and macro-economic decisions at the national level will inform social and political structures at the national, regional and local levels. The value of a political economy approach for such research is that while giving significance to economic processes it recognizes that existing social and political structures matter greatly.

The chapter is divided into three sections. The first relates Latin American industrialization to major shifts in the global economy during the twentieth century and to broad changes in macro-economic policy by Latin American governments. The second section describes some of the major institutional and technological characteristics of industrialization in Latin America. The third section tries to relate the process of industrialization in Latin America to that of urbanization. It has the starting point that compared with the cases of Western Europe, North America and East Asia, the linkages between industrial and urban growth are much less intimate in Latin America – but are nevertheless present.

It is necessary to make some definitions and to put the industrial category into the wider context of economic development. According to international statistical surveys, the 'industry' sector includes: mining; manufacturing; construction; and electricity, water and gas. Thus, industrial growth is more substantial than just that of the manufacturing sector. This is also reflected in Table 9.1, which ranks Latin American countries according to GNP per capita and identifies the distribution of GDP between industry, agriculture and services, but also includes manufacturing GDP as a sub-category of industry. It is interesting to contrast GNP per capita and the relationship between manufacturing GDP and agricultural GDP. In five of the six top-ranking countries, manufacturing GDP is more than twice that of agricultural GDP. Meanwhile, in the four countries with lowest GNP per capita, the agricultural GDP remains substantially higher than that of manufacturing. In this sense, then, the proportionate size of the manufacturing GDP would appear positively related to per capita income. The data can also be used to introduce the proposition that a dynamic manufacturing sector is vitally important for sustained economic growth in developing countries. Some East Asian countries have forcefully demonstrated this relationship in the latter half of the twentieth century. To what extent is this true in Latin America?

Table 9.1 Latin America: changes in the distribution of GDP, 1970–91 (per cent)

Country	Industry		Manufacturing*		Agriculture		Services		GNP per capita (US$)
	1970	1991	1970	1991	1970	1991	1970	1991	1991
Mexico	29	30	22	22	12	9	59	61	3,030
Brazil	38	39	29	26	12	10	49	51	2,940
Uruguay	37	32		25	19	10	44	58	2,840
Argentina	38	40	27		13	15	49	46	2,790
Venezuela	39	47	16	17	6	5	54	48	2,730
Chile†	41	36	26	21	7	9	52	55	2,160
Panama	21	11	12		15	10	64	79	2,130
Costa Rica	24	25		19	23	18	53	56	1,850
Paraguay	21	24	17	18	32	22	47	54	1,270
Colombia	28	35	21	20	25	17	47	48	1,260
El Salvador	23	24	19	19	28	10	48	66	1,080
Peru‡	32	37	20	20	19	14	50	49	1,070
Ecuador	25	35	18	21	24	15	51	50	1,000
Guatemala		20				26		55	930
Bolivia‡	32	29	13	15	20	16	48	55	650
Honduras	22	27	14	16	32	22	45	51	580
Nicaragua	26	23	20	19	25	30	49	47	460

* Manufacturing GDP is a sub-category of industry GDP
† Chilean Central Bank figures for 1991
‡ IADB figures for 1992

Main source: World Bank data

Latin American industry and the world economy

Latin American academics and 'Latin Americanist' social scientists have long seen the need to analyse the continent's economic growth in terms of its relations with the wider world economy. Furthermore, Latin American economies have been badly affected by two crises in the world economy in the twentieth century. Both crises had strong negative effects on production and both were followed by significant changes in policy by most Latin American governments. The first crisis was that of the World Depression, which started in 1929; the second has been called the debt crisis with its immediate origins in the events of 1982. Both crises had their impact on Latin American governments and their policies. The Depression caused most governments to become more inward-looking and suspicious of the world economy. The debt crisis, in contrast, caused some governments to seek solutions to the crisis through closer integration with the world economy. When it was these countries that emerged with the strongest economies towards the end of the 1980s, most other Latin American governments shifted to what became known as more 'outward-oriented' approaches.

The Depression and its aftermath

Before 1929, Latin American governments had generally supported free trade and close integration with the world economy. Economic growth in Latin America in the nineteenth century had greatly benefited from the rapid and substantial expansion in world trade and from the increasing flow of investment capital from such core countries as Britain. Latin America exported raw materials (mineral and agricultural) and imported manufactured goods. Most Latin American governments promoted free trade and export-led growth, at least until 1914, when the outbreak of the First World War ended 'the pristine epoch of free trade'. The First World War and its aftermath brought a shift to protectionism in the core countries of the world economy and Latin American governments reacted by modestly increasing tariffs and promoting industrial investment for domestic markets. However, it was not until the Depression of 1929–33 and the more than halving of world trade that Latin American governments faced the dire problems of a collapse in prices, export production and related employment. Between 1928 and 1933, Latin American exports declined from about $5,000 to $1,500 million. After an initial period in which governments vainly continued to boost exports, the reaction became one of delinking economies from that of the world economy. The increasing autarky of Latin American economies was the result.

Latin America changed from a set of free-trade economies to one of highly protected economies. Tariffs, quotas and exchange controls provided protection from foreign competitors by making the entry of foreign goods expensive or impossible. Latin American entrepreneurs, observing the scarcity of goods and the level of protection, began to produce or increase the production of goods previously imported. Industrial production and employment increased as a result. In Chile, manufacturing employment increased by over 5 per cent a year between 1928 and 1937; in sectors particularly favoured by protection, such as textiles, employment rose by an average of 16 per cent a year. By the end of the 1930s, Chile had become virtually self-sufficient in textiles.

The promotion of manufacturing behind high protective tariffs continued to be followed in most Latin American countries after the adverse effects of the Depression had diminished. It was argued that all developed countries had industrialized behind high protective tariffs. It was only after a country had developed a more mature industrial structure that it could become involved in the free trading of goods. In order to promote a more mature industrial structure, governments should actively intervene not only through the elaboration of industrial policy but also through the creation of development corporations (such as CORFO in Chile). Governments drew up strategic plans for industrial sectors and facilitated investment in key feedstock industries, such as steel, where private investors might be unwilling to venture.

Import substitution industrialization

A policy of import substitution industrialization (ISI) became formalized during the 1940s and 1950s. The policy broadly envisaged four overlapping phases of industrial production; governments would assist in each of these phases and in the progression from one phase to the next. The first phase had been started in the 1930s when tariffs, import quotas and multiple exchange rates had prevented or seriously hampered the import of those products deemed 'easy' to manufacture nationally – such as textiles, clothing, footwear, food products, pharmaceuticals and basic chemicals. The second stage of ISI became the focus of industrial policy in the late 1950s and 1960s. In the jargon of the time, this phase of industrial growth was to concentrate on 'consumer-durable products', such as refrigerators, cookers, televisions, motor vehicles.

The third stage was intimately linked with the second in that it had to promote 'feedstock' or 'intermediate' industries which would produce the inputs for 'second-stage' companies. 'Third-stage' industries were of two varieties. First, there were basic feedstock industries, such as steel, chemicals, petrochemicals, synthetic fibres, plastics and aluminium; many of these industries were developed through government-owned corporations. Secondly, there were component plants, producing the wide variety of parts needed by the second-stage industries. In this latter case, it was the role of government policy to make the formal link between second- and third-stage industries. One classic case was Kubitschek's *Programa de Metas* (1956–60) in Brazil. Within the industrial section of this programme, for example, motor vehicle production was to rise from negligible proportions in 1956 to about 350,000 vehicles in 1960. At the same time, there was to be a progressive 'nationalization' of supply so that between 1956 and 1960, Brazilian-made components in the vehicle assembly industry had to rise from 35 to 90 per cent of the total component cost of a vehicle. After completion of the second and third stages of ISI, a final stage would promote the development of the capital goods industry (machinery and plant installation). However, it was only in Brazil that a significant capital goods industry developed.

The high level of protection, particularly for the linked second and third stages, was partly justified by the 'infant industry' argument. As government industrial policy attempted to introduce new manu-facturing sectors into Latin American countries, it was argued that national firms needed time to learn new technologies and new manufacturing processes before the impact of global competition was felt. As policies of protection were implemented in new manufacturing sectors in Latin America, this argument had, in retrospect, two flaws. First, and unlike the East Asian experience, there were no plans for the phasing out of the high levels of protection. Secondly, as in the case of the Brazilian motor vehicle industry, it was multinational corporations rather than national corporations that benefited from trade protection.

This latter trend brought in its wake significant conflicts between Latin American governments and multinationals. For example, in Mexico, the relationship between government and vehicle multi-nationals during the 1970s became characterized by hard bargaining and increasing conflict. The Mexican government placed increasing demands on the multinationals that produced in Mexico – and were thus seen to enjoy supplying a protected market. These included higher levels of component supply from national sources (over 90 per cent) and the necessity for each multinational to achieve a trading surplus (exporting more vehicles and components in value terms than importing). The vehicle multinationals found these demands impossible to meet in the oil boom years of the late 1970s (it proved difficult even to supply the booming domestic market for vehicles) and hence serious conflicts emerged between the Mexican government and the vehicle multi-nationals.

In reality, the success of inward-oriented industrialization partly depended on the size of the national market. In large markets, such as that of Brazil, where competition between producers occurred and where reasonable economies of scale could be achieved, manufacturing industry could approach international levels of competitiveness. Indeed, as in the case of Brazilian industry in the 1970s, the provision of substantial subsidies for export by government could encourage producers to increase manufacturing exports rapidly (see Table 9.2). Thus, large countries recorded higher rates of manufacturing growth than smaller countries, particularly in the 1965–73 period when average growth rates of 12 per cent in Brazil and 8 per cent in Mexico were achieved. As a proportion of Latin American manufacturing production, Brazil and Mexico increased their combined weight from 43 per cent in 1950 to 62 per cent by 1978. The policy of ISI thus had the result of spatially concentrating manufacturing in just two Latin American countries.

In small markets, the low-cost production of consumer products was seriously hampered by a lack of economies of scale and competition between firms. Manufacturing production became high cost and suffered low internal demand. In times of stagnation, falling demand could start a vicious downward spiral of lower economies of scale, higher costs and even lower demand. With such high-cost production, there were great difficulties in ever expanding the potential market through exports.

For smaller countries, one much-heralded solution to the restrictions of low demand and low economies of scale was for countries to group

Table 9.2 Structure of exports in Latin America's ten most populous countries, 1965–82 (percentage share of merchandise exports)

	Fuels, minerals and metals		Other primary commodities		Manufactures	
	1965	**1982**	**1965**	**1982**	**1965**	**1982**
Dutch disease impact						
Venezuela	97	97	1	0	2	3
Mexico	22	78	62	10	17	12
Chile	89	65	7	27	5	8
Ecuador	2	64	96	33	2	3
Peru	45	69	54	17	1	14
Bolivia*	93	86	3	11	4	3
No Dutch disease impact						
Argentina	1	9	93	67	6	24
Brazil	9	18	83	43	8	39
Colombia	18	8	75	68	7	24
Guatemala	0	2	86	69	14	29

* 1979 not 1982 figures

Source: World Bank

together and form, through a process of economic integration, a much larger market (Plate 9.1). The Central American Common Market (CACM) created in 1960 and consisting of Guatemala, El Salvador, Nicaragua, Costa Rica and Honduras gave manufacturers a market of 10 million, up to ten times larger than previous national markets. In 1969, five Andean countries (Chile, Bolivia, Peru, Ecuador and Colombia) formed the Andean Pact. The prospect of an enlarged market of 70 million attracted all five countries and later Venezuela in 1973. In continental terms, however, schemes attempting to integrate countries adopting ISI strategies had little impact. The proportion of Latin American manufacturing in the twelve smaller countries involved in schemes of economic integration fell between 1950 and 1978 – from 22 to 20 per cent. Such schemes of integration were bedevilled by problems of trade diversion (low-cost supply replaced by high-cost supply), slow progress on intra-regional free trade and the spatial concentration of economic benefits in the more prosperous of the 'small' countries (Gwynne 1985).

Mineral economies and Dutch disease

The inward orientation of Latin American manufacturing was reinforced by the reliance of many countries on mineral or fuel exports. Table 9.2 demonstrates the structure of exports in Latin America's ten largest countries (according to population size) in 1965 and 1982. Six of the ten countries (Venezuela, Mexico, Chile, Ecuador, Peru and Bolivia) had

221

Plate 9.1 Electronic assembly plants in Tijuana, North Mexico. Squatter settlements lie in the foreground; public housing project houses in the background. Tijuana's local government has a serious lack of resources, hence large numbers of those employed in the factories have to build their own houses. The workers come from other parts of Mexico and their homes lack services, at least at first. (Photo: R. Gwynne)

more than 65 per cent of their 1982 exports consisting of fuels, minerals or metals.

Mineral economies include two main categories, the hydrocarbon producers (such as Venezuela, Mexico and Ecuador) and the hard mineral exporters, such as Chile (copper), Bolivia (tin) and Peru (copper, zinc). In the latter half of the twentieth century, mineral economies have recorded lower levels of economic growth and social welfare than non-mineral economies at a similar level of development, a finding that Auty (1993) finds counter-intuitive. The roots of the mineral economies' under-performance lie in the mining sector's production function (ratio of capital to labour), limited linkages with the national economy and deployment of rents. Mineral production is strongly capital-intensive, employs a very small fraction of the total national workforce and has historically required large inputs of capital from foreign sources. Thus, a large fraction of export earnings flows overseas to service the foreign capital investment. In this way, fiscal linkages (through taxes paid by mining corporations to national governments) tend to dominate the mining sector's contribution to the national economy.

The poor performance of mineral economies also lies in the negative impact they have on the agricultural and manufacturing sectors. Mineral economies tend to produce substantial rents – revenues in excess of production costs and a normal return on capital (profit). Since the early 1970s and the nationalization of most of the remaining foreign-owned mineral resources (in Venezuela, Chile and Peru for example), more rents have been captured by government through taxation and corporate ownership. In these circumstances, a process known as 'Dutch disease' can occur, particularly during periods when international commodity prices are booming, as in the mid-1970s and early 1980s (see Box 9.1).

Box 9.1: Dutch disease

Dutch disease results from a strengthening (appreciation) of the exchange rate as a result of rapid increases in the inflow of mineral rents into the domestic economy. Such increases can result from a boom in commodity prices and/or from rapidly increasing production. An overvalued exchange rate invariably results, signifying that exports become expensive but imports cheap. Mineral and fuel exports remain buoyant due to high international prices for these commodities, but agricultural and manufacturing exports, which are much more price-sensitive, cannot compete in international markets.

Within inward-oriented countries, Dutch disease merely intensifies the levels of protection and the autarky of what economists call non-mining tradeables (agriculture and manufacturing). Agriculture and manufacturing become even more closely wedded to the national market. The real problem for the mineral economies emerges when international mineral prices begin to fall in the cyclical downswing, as for example occurred in the mid-1980s. The value of mineral exports falls significantly, causing major problems of external financing – particularly if the mineral economy spent rather than saved its windfall receipts during the boom years (as normally occurred). The mineral economy cannot readily diversify its export base into agricultural or manufacturing sectors – as these were left internationally uncompetitive and inward-oriented. As a result, in Latin America, the policy of ISI has been reinforced by the fact that many (though not all) countries are mineral economies in which overvalued exchange rates during cyclical commodity booms cause an accentuated inward orientation of manufacturing. This is the basis of the resource curse thesis – that substantial

223

endowments of non-renewable resources (mineral and hydrocarbon) can be counter-productive to the aim of a developing country for sustained economic growth.

The sustainable development of mineral economies must lie in successful diversification into competitive non-mining tradeables: agriculture, forestry, fishing and manufacturing. During the 1980s, Chile was able to diversify its export base away from a dependence on copper; by 1993, fruit, forest products and fish exports accounted for over one-half of Chilean exports. Furthermore, the management of copper price volatility has been substantially improved by the creation of the Copper Stabilization Fund in 1987. This innovative fund, controlled by the Central Bank, means that windfall receipts in the period when the copper price is climbing are saved to supplement government expenditure in the cyclical downswing of the copper price. The Chilean case shows that diversification of exports and stricter management of windfall receipts permits a mineral economy to become more outward-oriented and reduce the impact of overvalued exchange rates during commodity price booms; however, even in Chile, the exchange rate became more overvalued during the mid-1990s.

The debt crisis and the shift to open economy policies

In the 1970s, both the more successful manufacturing economies (Brazil, Mexico) and mineral economies (Chile, Venezuela, Ecuador, Peru) became the recipients of loans from international commercial banks. Such loans operated to reinforce inward orientation in the late 1970s, particularly in the rich oil economies (Mexico, Venezuela, Ecuador, Peru). It was when the government of one of these countries, Mexico, declared a moratorium on debt repayments in August 1982 that the debt crisis began. However, the debt crisis and its aftermath proved the catalyst for Latin American countries to move to more outward-oriented policies.

Why did the debt crisis cause many Latin American countries to turn to more free market open economy policies (FMOEPs)? There are at least three reasons. First, the debt crisis identified the economically unsustainable nature of ISI and a narrow range of exports. In 1982, world trade was in decline at a time when real interest rates were very high throughout the world. The external accounts of all Latin American countries were in serious disarray – exports declining in both value and quantity, interest payments on loans soaring, national debt rapidly increasing without any sign of how its growth would stop – unless economic policy changes occurred.

Second, the shift to FMOEPs was strongly recommended by international organizations, such as the International Monetary Fund (IMF), the World Bank and the Inter-American Development Bank. They argued that FMOEPs fit in better with the realities of the modern world economy. The IMF in particular favoured structural adjustment policies which involved seven aspects: macroeconomic stability, deregulation, privatization, openness to trade, financial efficiency, a poverty pro-

224

gramme and solid institutions. Such policies were more realistic in terms of the progressive internationalization of the world economy which makes national autarky a difficult proposition.

Third, it was precisely those countries which quickly turned to FMOEPs after the debt crisis (Chile and Mexico) that subsequently showed signs of rapid economic growth in the late 1980s. Chile had shifted to more free market policies in the mid-1970s but in 1979 its economic ministers retracted from the policy by fixing the Chilean peso to the US dollar. The consequent economic collapse and debt crisis of 1982/83 had the impact of strengthening free market open economy policies in Chile; between 1986 and 1995, Chilean economic growth averaged 6 per cent per annum. In a similar vein, the origins of structural adjustment in Mexico can be directly traced back to the need to do something after the beginning of the external debt crisis of 1982. The comparative success of both the Mexican and Chilean economies in the late 1980s gave a clear sign to other Latin American governments of the virtues of FMOEPs.

The shift to FMOEPs had two quite logical but contrasting results in terms of sectoral development. First, there was a significant and immediate decline in production and employment of inward-oriented industry. Second, but after a certain delay, a diversified growth in exports occurred, either based on primary products (as in Chile) or on manufacturing (as in Mexico). The immediate decline of one set of economic sectors that is not internationally competitive and the delayed growth of a set of sectors that is able to operate successfully in global markets meant that the shift to FMOEPs involved significant short-term difficulties. Delayed growth of export-oriented sectors occurs because it takes time to build up investment and confidence, change business practices and forge competitiveness in sectors that can come to operate successfully in global markets. In the short term, then, national unemployment rates soar. In Chile, they rose above 20 per cent in the mid-1970s and above 30 per cent in 1983/84. However, by 1989, with the achievement of a sustained outward-oriented economic growth, Chilean unemployment had fallen to 6 per cent. Such huge shifts in unemployment caused major social and political tensions in Chile, particularly in the 1983/84 period. In this context, it is important to stress that in 1983/84 Chile was governed by a military dictatorship willing to engage in severe repression of popular protest by the impoverished sections of the community.

In some countries, the short-term difficulties caused in shifting from ISI to FMOEP has caused a reversal of policy (as in Venezuela in 1993/94) or a significant moderation in the scale and scope of structural adjustment policies (Brazil). However, the shift from ISI to FMOEP does not have to be linked to an authoritarian regime. In both Mexico and Argentina, the shift has taken place within a democratic framework. However, in both cases governments have been aware of the need to explain their policies clearly to the electorate.

Renewed attempts at economic integration

During the 1980s, when most of Latin America was in the economic doldrums, FMOEPs implied individual countries boosting trade with developed countries – particularly the USA, the EC and Japan. In order to increase inward investment, links with firms in developed countries were nurtured. In this way, most of the economies of Latin American countries became distinctly more multilateral in the global context, with firms promoting trading and investment links with a wide variety of developed and newly industrializing countries.

It is within this context of increasing multilateralism that a renewed interest in economic integration is taking place. There is an apparent paradox here. The 1980s have shown that for countries to extricate themselves from the external constraints of the debt crisis, they must forge stronger trading, investment and technology links with developed countries – as the Chilean case demonstrates. However, the political emphasis in the 1990s on Latin American countries integrating within regional trading groups indicates a different strategy and set of economic priorities. Many Latin American governments perceive that advanced countries are closing off their markets to the exports of developing countries. The creation of a single European market in 1992 was certainly seen in this light by many Latin American politicians and business people. In general, they see trade policy in advanced countries shifting away from multilateralism to increasing regionalism.

The countries of mainland Latin America can be divided in two according to government integration policy (see Fig. 9.1):

1. Those committed to promoting economic integration with a developed country (USA) rather than Latin American economic integration – Mexico (through the North American Free Trade Agreement, NAFTA) and Chile (due to become a NAFTA member in 1996).
2. Those committed to regional economic integration rather than close links with the USA. Three regional trade groupings are evident:
    (a) Central American Common Market (CACM) – Guatemala, El Salvador, Honduras, Nicaragua, Costa Rica;
    (b) Andean Group – Venezuela, Colombia, Ecuador, Peru, Bolivia;
    (c) MERCOSUR – Argentina, Brazil, Paraguay, Uruguay

What are the most likely trends in Latin American economic integration during the late 1990s? Two, somewhat conflicting, processes seem likely to occur. First, the US government, whilst maintaining its broad allegiance to multilateralism, will gradually expand regional free trade agreements in Latin America. US Congressional approval will be increasingly slow and painstaking. But the process will be justified in the USA as one that encourages the extension of free trade in Latin America. Secondly, there should be gradual progress in the evolution of Latin American regional free trade areas – CACM, the Andean Group, MERCOSUR. The more realistic attitude towards freer trade and inward investment in these integration schemes will mean that many of the

Fig. 9.1 Economic integration in Latin America

mistakes of the 1960s will not be repeated. However, increasing economic integration will not be a harmonious process. Tensions will emerge between countries as benefits are seen to be asymmetrically distributed within the member countries of the integration schemes. However, the political imperative of the need for closer regional integration within a world economy increasingly divided into trading blocs should mean that Latin American governments will be more willing to bury their differences and disputes than they were in the 1960s.

227

Major characteristics of manufacturing in Latin America

Technological change

Under ISI, there evolved a distinctive spatial concentration of the production of intermediate, consumer and capital goods in Brazil, Mexico and Argentina. This was linked not only with size of market but also with the internal generation and adaptation of new technology. Unlike the situation in the more advanced countries, much of the process of technological change which takes place in Latin America consists of the imitation of products and processes that have already been developed in more advanced countries. At this stage of purchasing technological designs, there are few differences among Latin American countries. It is at the next stage, the adaptation and improvement of imported technology to meet local circumstances, that the larger countries had considerable advantages over other countries under ISI. Argentina, Brazil and Mexico had a substantial number of highly trained professional and technical personnel able to adapt imported technology to local conditions and to generate national technological knowledge.

However, under ISI, such technology generation was often not particularly cost-effective. According to Katz's (1987) detailed study, 'cost reduction was not necessarily a priority of the technological search efforts undertaken by Latin American firms'. He demonstrated that under ISI manufacturing firms in Latin America settled for 'discontinuous technology' rather than continuous flow and highly automated technologies. Compared with the contemporary development of technological capability in East Asia, Latin American firms were notably poor in developing process technology. Factories frequently produced goods or services in small runs or in response to individual orders. The plant was not designed following the array of technical transformations demanded by one specific product, but rather by 'groups' of somewhat similar machines or tasks. There were then huge differences between the highly automated but flexible process technologies being developed in East Asia at the time and the costly, archaic and time-consuming process technologies characteristic of Latin American firms. The difference can partly be explained by the fact that East Asian industrial firms were intent on carving out large international markets (and invested heavily in technology to reduce unit costs) whilst Latin American companies had their horizons restricted to the national market-place; in these markets, monopoly or oligopoly power meant that cost-effectiveness was not seen as a high priority for the generation of new technology.

This contrast between East Asian and Latin American firms has been maintained even after Latin American governments shifted to FMOEP in the late 1980s. The question for 'new exporters' can be one of appropriate (and normally more specialized) technology. Using the East Asian experience as a guide, there is an important role for the state in promoting and fostering 'national technological capability'. However, in Latin America, those governments that have shifted to FMOEP have done so against an ideological backdrop of *'laissez-faire'*. With Latin American governments trying to distance themselves from intervention in industrial policy there has been little enthusiasm for drawing up policies to develop national technological capability.

The institutional framework: state firms and multinationals

Until the 1980s, the process of industrial expansion in Latin America had been engineered through a distinctive institutional structure. Under ISI, the institutional structure was referred to as the triple alliance: between state firms, multinational corporations and national private enterprise. Within the latter category an important distinction can be made between large- and medium-sized firms on the one hand and small firms on the other.

State firms had become dominant in the extractive industries and in the further processing and refining of hydrocarbons and minerals. State firms developed in these sectors were linked to political objectives which stressed the need for the nation to exert greater control over the non-renewable resources of the country, resources that were crucial for exports, taxes, the fiscal budget, employment and the exchange rate. Furthermore, under ISI, it was the role of state enterprise to invest in those intermediate industries that continued industrial expansion required. It was argued that as the investment was often of a strategic nature (power, communications, steel), governments should exclude the participation of multinationals. Furthermore, in terms of national enterprise, it was only the state that could provide the necessary capital for such large investments. Since the shift to FMOEP, many of these enterprises have been privatized. For example, in Chile, the latter half of the 1980s saw the privatization of such strategic sectors as steel, electricity generation and distribution, coal production, telecommunications and air transport (though not copper mining). Most of the shares in these companies were bought up by private national investors (particularly pension funds) but in telecommunications, foreign interests have taken majority control.

The multinational enterprise in Latin America has historically been of Western European and US origin but Japanese interests have been increasing recently, particularly in Mexico. The role of multinational enterprise as an agent of Latin American industrialization has been distinctive, not least because no other continent has received such an important contribution from foreign companies in their process of industrialization. The USA, Japan and the countries of Western Europe largely industrialized through national institutions, whether state or private. Other developing countries are presently industrializing with the strong participation of multinational enterprise, but Latin American countries have already developed complex and diverse industrial structures from the contributions of foreign investment. Multinational investment has been particularly significant in those two countries which achieved relatively mature industrial structures under ISI, Brazil and Mexico; over two-thirds of direct foreign investment (DFI) in Latin American manufacturing was concentrated in these two countries in the late 1970s, over one-half coming from US companies.

Under ISI, multinational investment tended to concentrate in the more dynamic and technologically innovative sectors. In the mid-1970s, over 75 per cent of DFI in Brazil was channelled into such sectors as vehicles, machine tools, pharmaceuticals, chemicals, communications, electrical and medical industries; in 1979 multinational enterprise controlled 56 per cent of assets in the transport sector and 51 per cent in the electrical. Multinationals dominated in these sectors because they had direct access to the latest technology and because there was competition between multinationals to be present in such large but closed markets as Brazil and Mexico. With the shift to more free-market and open-economy policies in Latin America, direct foreign investment is increasingly being encouraged. As a result, the role of multinationals in the manufacturing sector is likely to become more and more significant.

Multinationals and the shift to outward orientation in Mexico

With the shift to FMOEP in Mexico since 1982, there has been a considerable increase in inward investment from multinational companies. Multinationals are attracted to those countries that have deregulated their economies; they are in need of cheap labour and an open productive environment in which firms are free to make rapid adaptations in production (such as product types or quantities of labour employed) in order to retain international competitiveness. However, in any shift to outward orientation, Mexico's economy becomes more intertwined with the world's strongest economy, that of the USA. The relationship is a distinctly unequal one. As Table 9.3 demonstrates, Mexico's exports to the USA constituted nearly 75 per cent of its total exports in 1991; US exports to Mexico were only about 8 per cent of total exports.

The signing of the NAFTA between the USA, Canada and Mexico in 1992 (and its subsequent ratification by the US Congress in 1993) has been a powerful magnet for multinational investment in Mexico – both prior to the signing and subsequently. In this context, the relationship between Mexico and the USA can be viewed within the classic framework of multinationals and the international division of labour. In the early 1990s wages in Mexico were about one-tenth of those north of the border. Multinationals have therefore specialized in locating labour-intensive activities south of the border (whilst maintaining more capital-intensive and technologically innovative plant north of the border). Manufacturing trade between Mexico and the USA is thus becoming increasingly characterized by a northward flow of labour-intensive goods and a southward flow of technologically advanced products (particularly capital goods and machinery). The corollary for technological development is that the USA will continue to be the

Table 9.3 Trilateral trade patterns among the NAFTA nations (percentage of total trade)

| | | 1980 | | 1991 | |
		Imports	Exports	Imports	Exports
Canada	USA	67.5	60.6	61.9	75.1
	Mexico	0.5	0.6	1.6	0.3
USA	Canada	16.3	16.0	18.4	20.1
	Mexico	5.0	4.5	6.2	7.9
Mexico	Canada	1.7	0.7	0.8	5.4
	USA	58.8	64.7	70.7	74.5

Source: International Monetary Fund

generating country and Mexico the receiving country for new technology.

According to Kenney and Florida (1994), in such a pattern:

> Mexico attracts plants which have production processes that require cheap labor and are in markets suffering from intense competition. It is highly unlikely that the more capital-intensive production processes that rely on high levels of skill formation among workers will be relocated to Mexico.

The contrast between the type of industry in Mexico and the USA is most evident within the towns of the Mexican border industrialization programme (see Fig 9.2). Here a manufacturing plant can import duty-free all those raw materials and parts to be assembled and re-exported to the USA; when the finished goods enter the USA, duty is paid only on the added value.

Multinationals from different countries have developed contrasting strategies to plant location in the Mexican border area. US multinationals very much see it as a location for the production of both labour-intensive products and components; labour-intensive components are used in the more capital-intensive assembly plants in the USA. There is evidence that US multinationals are shifting investment from Asia to Mexico; for example, Zenith has decided to close its facilities in Taiwan and in the USA and relocate all of its television production in Mexico.

Japanese corporations, meanwhile, see the location of a *maquila* (assembly plant) in a Mexican border town as part of their international production chain, in which the final product is geared to the US market. The *maquilas* actually produce few of the components of the final product. They essentially assemble advanced, high-quality components which are manufactured in plants in Japan, South-East Asia and the USA. According to Kenney and Florida (1994), 'such activities remain extremely sensitive to labor costs, can be effectively performed with unskilled labor, and are sufficiently standard so as to not require continuous improvement activity'.

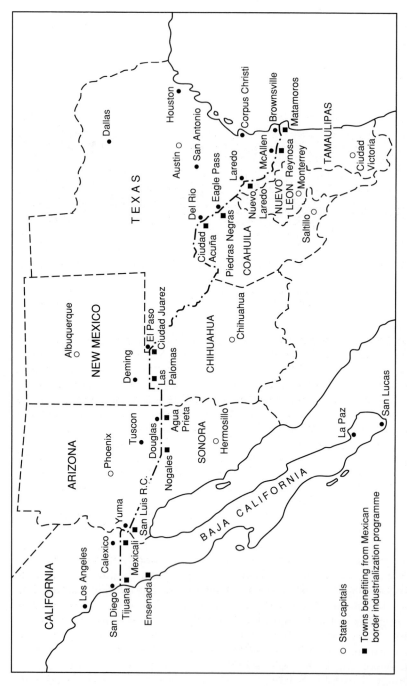

Fig. 9.2 The Mexican Border Industrialization Programme *Source:* Gwynne (1990)

Plate 9.2   The Sony factory of Tijuana. Tijuana is a typical cheap labour
location with direct access to the US market. (Photo: R. Gwynne)

The Mexican border therefore provides a landscape dominated by
multinational capital, where international linkages rather than local
linkages predominate; Mexican suppliers are notable by their absence in
the border towns. However, the spatial impact of multinational investment in Mexico goes far beyond the relatively narrow confines of the
border programme. Figure 9.3 reveals the spatial evolution of Mexico's
vehicle sector, dominated (at least in assembly and engine production)
by foreign multinationals – three US corporations (GM, Ford, Chrysler),
two European (VW, Renault) and one Japanese (Nissan). During ISI, all
corporations chose assembly locations either in Mexico City or in
neighbouring towns. During the late 1970s and early 1980s, the Mexican
government forced the vehicle multinationals to invest in component
manufacture for export (mainly engines) so that each corporation would
balance its exports and imports. As an important part of engine
production would be for export, the vehicle corporations located their
engine plants further north – with the exception of VW. GM and
Chrysler set up plants at Ramos Arizpe (near Saltillo), Nissan at
Aguascalientes, Renault at Gomez Palacio and Ford at Chihuahua. The
single main export market for GM, Chrysler, Ford and Nissan engines
was the USA and the respective assembly plants of these corporations
(the main markets for Renault and VW engines were European and
Latin American). By the mid-1980s, Mexico was the source of about 40
per cent of total engine imports into the United States.

After the shift of Mexico to FMOEP in the mid-1980s, the vehicle
multinationals saw the potential for locating assembly plants in Mexico,
geared to the US market. The Ford US$500 million assembly plant in
Hermosillo is a highly specialized plant closely integrated with the US

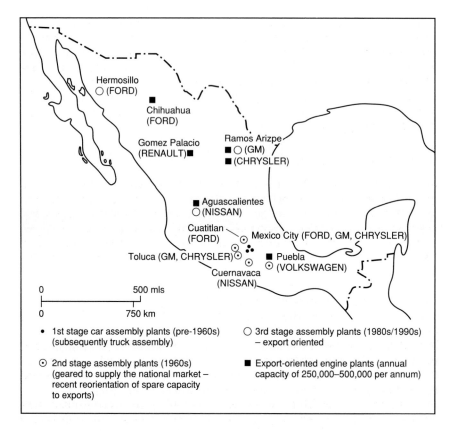

Fig. 9.3 The spatial evolution of the Mexican motor vehicle industry

economy and market. Opening in November 1986, it produced the sub-compact Tracer model (specially designed for the US market and based on the Ford Escort), using process technology designed by Ford's Japanese partner, Mazda. Most production (170,000 vehicles per annum by 1990) was exported to the USA and the integration of the Mexican plant with US suppliers also became close: by 1990, component producers in the USA provided 60 per cent of supplies with the remaining 40 per cent coming from Japan and Mexico.

GM and Nissan have also built export-oriented, technologically advanced plants since 1980. Much of GM's production from its Ramos Arizpe plant is exported to the USA. In contrast, Nissan uses its Mexican assembly plants to supply Latin America (the US market is mainly supplied from Nissan's Smyrna plant in Tennessee). Two other vehicle multinationals, VW and Chrysler, modernized their old plants in Puebla and Toluca respectively; their major market is still Mexico, but vehicle exports have increased.

The vehicle sector has become one of the more technologically advanced in Mexico in terms of both product and process technology.

Vehicles and components are now one of the main manufacturing export categories of Mexico. However, such international prominence and technological sophistication must be explained in terms of the close technological, supply, financial and marketing linkages that have evolved with both Japanese and US multinationals. The Latin American characteristic of high-technology industry being closely linked to the multinational corporations has intensified with FMOEP in the larger manufacturing countries, and particularly in Mexico.

National private enterprise

In contrast to the state firm and multinational corporation, national private enterprise is characterized by great diversity in terms of size, technological level and forms of organization. In most large and medium-sized countries, large national conglomerates have developed with a wide variety of manufacturing interests and related services in such areas as banking, insurance, finance, tourism, commerce and the media.

Large- and medium-scale national firms in Latin America have been the focus of limited research, at least compared with multinationals and small firms. Yet it is these firms that are experiencing the greatest pressures to restructure in the 1990s. After half a century of protection behind tariff walls, these firms are now having to vigorously defend their national market shares from cheap imported products and attempt to expand into export markets. The task is too great for many firms and hence substantial reorganization is taking place in the sector.

One particular problem is that of firms having to revolutionize their business culture. After decades of being accustomed to supply the limited horizons of the national market, national firms, in order to survive, have now to change their traditional strategies and seek to supply international markets. Messner (1993) highlights some of the problems that firms in the Chilean wood products industry have faced in making such a transition:

> The most serious problem the firms initially faced was recognising the completely different requirements of the world market compared with those of the domestic market and translating them into appropriate corporate strategies. It was particularly important to improve such non-price aspects of competition as the organisation of work, the technological level of production, product quality, design and image, punctuality of deliveries, and marketing.

Most Chilean wood product firms were inward-oriented until the mid-1980s, had little contact with suppliers (autarkic mentality) and, because local demand was limited, offered a very wide range of products. Overcoming these entrepreneurial structures and strategies was not only a capital, know-how and technology problem but also a problem of business culture:

> in the early stages of reorientation the first generation of 'new exporters' were still convinced of the efficiency and viability of the entrepreneurial

235

concepts that had proved successful in the domestic market. Consequently, many initial attempts to export failed, and the continued application of past production concepts led to unwise investments (Messner, 1993).

Some firms had invested in new manufacturing equipment with a view to exporting a wide range of products, but on Messner's plant visit some of the equipment was standing idle.

However, some large and successful national conglomerates have emerged in most Latin American countries as restructuring has taken place. The more successful ones have been those that have changed their business culture, become more international in outlook and have expanded by taking over those firms that were unable to make the change. As a result, conglomerates are developing with investments in many different companies which themselves operate in many different sectors. However, at their base, many still rely on a close relationship with small enterprises – simply because these can often provide low-cost inputs for their wider operations.

Small firms

It is difficult to get a comprehensive picture of the quantitative and qualitative development of small firms over recent years. 'Small firms' and the 'informal sector' are elusive concepts, which hide a large heterogeneity in the types of firms. A clear and generally accepted definition of the informal sector or small firm is still lacking. In spite of almost 20 years of debate and research on the informal and small firm sector, it is not very well established in statistical terms. One useful classification of small firms in Peru has been produced by Villaran (1993). In marked contrast to structures in an advanced industrialized country, he divided the Peruvian industrial sector into five sub-sectors, three of which corresponded to small enterprise. These were:

1. Craft work: enterprises defined by the use of traditional craft technology and with an average capital density of US$300 of investment in fixed assets per job; 165,000 people were employed in craft work (22.9 per cent of the economically active population with 5 per cent of GNP in industry).
2. Micro-industry: characterized by the limited size of the firms (1–4 people employed) and low density of capital, equivalent to $600 per job; 210,000 people were employed (29.3 per cent of the economically active population with 8 per cent of GNP in industry).
3. Small-scale industry: firms ranged in size from 5 to 19 employees and had an average capital density of $3,000 per job; 137,000 people were employed (19 per cent of the economically active population with 13 per cent of GNP in industry).

Thus, in Peru, the three sub-sectors of small-scale industry employed 512,000 people in 1987; this was equivalent to 71.2 per cent of

Plate 9.3   Hat forming, Solano, Ecuador. Straw hats (Panama hats) provide
income and employment for suppliers of palm fibre from coastal
Ecuador, for dealers who contract hats to be worn and (here) for
retailers who shape hats according to prevailing fashion. (Photo:
D Preston)

employment in the industrial sector. Small-scale enterprise was thus a
major employer of labour but it was only responsible for 26 per cent of
the GNP. Villaran (1993) saw the craft and micro-industry sectors as
fully part of the 'informal' sector, although only 30 per cent of
small-scale industry could be so categorized. He saw firms in the small-
and medium-scale sectors as located in competitive markets. In contrast,
large-scale firms normally held positions of monopoly or oligopoly,
used capital-intensive imported technologies, charged high prices and
were generally uncompetitive in international markets – and would
have been in the Peruvian market if protection had come to an end.
However, in 1987, this sector was responsible for 46 per cent of
industrial GNP.

Many politicians and advisers in the 1990s argue that the shift to freer markets and more open economies must be linked to state deregulation and greater competition in such productive sectors as manufacturing. Deregulation can sound the death-knell for the more bloated and inefficient large-scale industries but provide a significant impetus for growth in the small- and medium-scale firms that had been previously restricted by uncompetitive practices and closed markets. This is what Schmitz (1993) calls the optimistic scenario, in which the growth of small producers is seen as open-ended, provided that the small producer has the drive and energy; evolutionary growth from small to medium to large is possible.

According to Schmitz (1993), 'even more than in advanced countries, competitiveness requires the capacity to adapt to disruptive circumstances'. Schmitz argues that during the debt crisis and its aftermath, as Latin American economies became increasingly more open, small firms adapted better than medium- and large-scale industry. The economic crisis of the early 1980s had a particularly negative effect on large firms whereas many small firms fared relatively well; the extensive use of subcontracted artisans enabled them to cope with volatile demand. The small firm sector is then a major employer of labour and possesses a somewhat greater flexibility than larger firms, particularly during the period of ISI. With the shift to FMOEP in the late 1980s and early 1990s it could be argued that large and medium-sized firms will also have to become more flexible and adaptable in order to survive.

Thus, in the big cities of Latin America, a large informal sector has developed alongside the formal sector of government activities and large and medium-sized firms. Small firms fall uneasily across this border. Some are flexible, efficient and resilient, able to adapt to economic crises better than large firms. The framework of labour surplus, via subcontracting networks, means that small firms can incorporate low wages and informal labour, thus reducing production costs significantly. Many small-scale enterprises, however, can best be seen as survival mechanisms, particularly when one refers to small-scale enterprises in service activities; even here, small enterprises may act to reduce costs of manufacturing industry (by providing recycled materials, such as paper, cheaply).

Industrialization and urbanization

The process of industrialization in Latin America cannot be understood without reference to the cyclical patterns of the world economy in the twentieth century. Before attempting to examine the links between industrialization and urbanization, it is first necessary to summarize some of the major characteristics of Latin American industry. First, and until very recently, industrial firms have supplied only the national market. Secondly, this inward-oriented industrialization has been associated with an emphasis on consumer good production. Such an orientation of production caused a third characteristic – a dependence on foreign technology. This reliance on foreign technology contributed to a fourth feature – a relatively small generation of employment given the importance of manufacturing to the economy (see Table 9.1), the size of the overall labour force and the large numbers of unemployed/underemployed. Indeed, the manufacturing workforce in the large and medium-sized firms in the formal sector, earning high and regular wages compared with their counterparts in the informal and non-manufacturing sectors, were often seen as a sort of labour aristocracy – relatively prosperous but small in number.

What implications has such a distinctive form of industrialization had on the process of urbanization? Two general points can be made. First, there has not been such an intimate connection between industrialization and urbanization as occurred in nineteenth century Europe or twentieth century East Asia (Japan, Taiwan, South Korea); urban growth in these contexts closely followed industrial growth and the consequent expansion of employment. In Latin America, however, increases in manufacturing employment have not been nearly so dramatic as the growth in urban dwellers. The growth in urban dwellers, caused by high urban fertility rates and large flows of migrants from rural areas, has far outweighed increases in manufacturing employment.

Secondly, inward-oriented industrialization reinforced the spatially concentrated pattern of urbanization in Latin America. Primacy became a distinctive feature of the urban geography of most Spanish American countries .during the post-colonial and export economy phases of the nineteenth century. For example, by the turn of the century both Peru and Argentina had primate cities with populations more than seven times as large as their second cities. However, by 1980, after 80 years of industrialization, Lima and Buenos Aires were more than 11 times larger than their respective second cities. Latin America tends to contain more primate cities than most other parts of the world. Only Bolivia, Brazil, Colombia, Ecuador and Honduras do not have a primate city whose population exceeds that of the second city of the country by at least three times. Even among the exceptions, Brazil, Ecuador and Honduras hardly count since they contain two cities both of which greatly exceed the population of the third city in the country.

In virtually all Spanish American countries (apart from Ecuador), the primate city and the political capital have been one and the same; Brazil is also an exception with the political capital in Brasilia (population of 2.36 million in 1990) and with São Paulo the primate city (population of 17.4 million in 1990). The evolution of national transport networks in the late nineteenth and twentieth centuries (both rail and road) have generally radiated out from the capital/primate city (Colombia is an exception owing to the evolution of regional transport systems). As a result, these cities became the most accessible locations in each country and a most advantageous location for at least two types of inward-oriented industry: consumer-good industries supplying the whole of the national market from one or two plants; and industries in which various raw materials from different locations in the periphery were required for a manufacturing process. The steady inflow of migrants maintained a labour surplus in the primate city and hence kept wages and labour costs down, especially for unskilled or semi-skilled labour.

The relationship between primacy and industrialization became mutually reinforcing. Industrialization increased employment in the primate city, both directly, in terms of the expansion of the manufacturing workforce, and indirectly, in terms of the growth of related employment in public and private services. Furthermore, the process of industrialization created diverse but powerful élites in the primate city – representing both national and international organizations and not only in industry but also in banking, commerce and political organizations. In Brazil, São Paulo's rapid industrial growth was linked to regional characteristics of entrepreneurialism – so strong that it emerged as the primate city of Brazil in the latter half of the twentieth century. In Colombia, the growth of regional manufacturing in Medellín and Cali restricted the national dominance of Bogotá. However, in Mexico, the growth of regional manufacturing in Monterrey did little to counter the national pre-eminence of Mexico City in manufacturing production and employment.

Schmitz (1993) points out that the primate city offers an attractive location not only for large and medium-sized firms but also for small firms. Schmitz takes the example of the weaving industry in São Paulo state in which small weaving firms predominate, particularly in Americana, and stresses the spatial linkages afforded by technology suppliers. Most loom manufacturers were located in São Paulo state and their proximity 'gave the users a greater chance of obtaining spare parts quickly, important because secondhand machinery was frequently used'. Where spare parts were no longer available from the manufacturers, small firms could fall back on secondhand dealers or on small engineering firms which could repair or copy a part. Thus, Schmitz (1993) concludes that small firms are attracted to primate city locations by the economies of agglomeration that 'arise when a network of suppliers develops that provides materials, tools, new machinery, secondhand machinery, spare parts, repair services and so on. Small firms individually cannot attain flexible specialization: it is the sectoral agglomeration which gives them their relative strength.' In such agglomerations, small firms benefit from collective efficiency and, in this context, elements of instability can be overcome more easily.

The spatial association of industrialization with the primate city has had the corollary that towns further down the national urban hierarchy have attracted little industry. There are some exceptions, such as Monterrey and Guadalajara in Mexico, Medellín and Cali in Colombia and Cordoba in Argentina. Inward-oriented, consumer-good industries were strongly attracted to the large city and avoided smaller provincial towns. The need for domestic firms to be near the major internal market and the locus of political power (important as the process of ISI was associated with strong government intervention) were vital locational considerations. Meanwhile, multinational firms were attracted to the cosmopolitan capital in which international communications were superior to alternative locations. As a result, manufacturing has had relatively little impact on the occupational structures of small and medium-sized towns in Latin America. Table 9.4 demonstrates this relationship for the Chilean urban hierarchy in 1970, when ISI was still firmly in place. Santiago and the port-industrial metropolitan area of Concepción had approximately one-quarter of their substantial workforces in the manufacturing sector. Meanwhile, small and medium-sized towns in other Chilean regions had only about one in eight of their workers employed in manufacturing. However, the attraction of a location near the primate city for industrial firms is reflected in the differential performance of small and medium-sized towns located in Santiago province at a short distance from Santiago; San Bernardo, Puente Alto and Peñaflor (see Fig. 9.4) had up to 35 per cent of their workforce involved in manufacturing in 1970. A similar pattern of small and medium-sized towns specializing in manufacturing near to the primate city can also be found near São Paulo (industrial towns such as Campinas, Piracicaba, Americana and Sorocaba), Mexico (Toluca, Cuernavaca, Puebla) and Caracas (Maracay, Valencia).

Table 9.4 Manufacturing workers as a percentage of the total non-agricultural workforce for urban areas in Chile, 1970

	Total non-agricultural workforce (A)	Manufacturing workforce (B)	B as % of A
Metropolitan areas			
Santiago	848,606	199,729	23.5
Concepción	113,014	28,727	25.4
Medium-sized towns in Chilean regions			
Antofagasta (North)	33,516	3,499	10.4
Temuco (South)	30,431	4,442	14.6
Rancagua (Central Valley)	24,682	2,389	9.7
Chillan (Central Valley)	23,691	3,035	12.8
Puerto Montt (South)	19,074	2,444	12.8
Los Angeles (South)	14,765	1,819	12.3
Coquimbo (North)	12,564	1,251	10.0
Curico (Central Valley)	12,172	1,353	11.1
Medium-sized towns in Santiago Province			
San Bernardo	31,504	7,069	22.4
Puente Alto	21,600	6,622	30.7
Peñaflor	8,638	3,050	35.3

Source: Chilean Census of 1970

The shift from ISI to outward orientation has had some small impact in reducing the dominance of the primate city in manufacturing employment. This is mainly because of the growth in resource-oriented

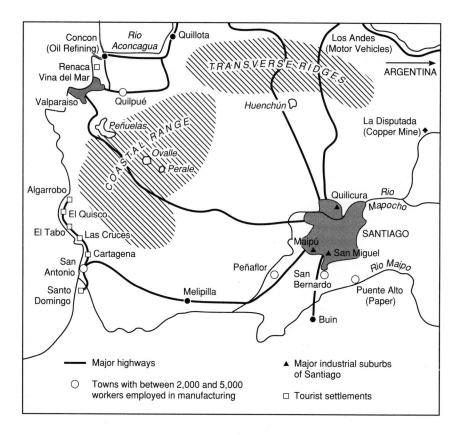

Fig. 9.4 Chilean central region

industries under outward orientation and the distribution of these resources in the periphery of Latin American countries. For example, the shift to outward orientation in Chile has been associated with a rapid growth of that country's forestry sector and substantial investments (both by multinational and national firms) in cellulose plants. These plants have been located within Chile's forestry region partly because of the high cost of transporting logs for long distances (see Fig. 9.5). Cellulose plants have generally not been located in the main market towns (except in Arauco) but in small towns (Nacimiento, Laja, San Pedro) or even villages (Mininco) due partly to the significant air pollution that is often associated with such plants. However, each plant has been based on imported technology, is capital-intensive and thus employs relatively little labour (an average of about 500 employees). Thus, although physical production may be increasing in the periphery, employment increases are much more modest.

Thus, the shift to outward orientation has not greatly changed the overall relationship between industrialization and urbanization as

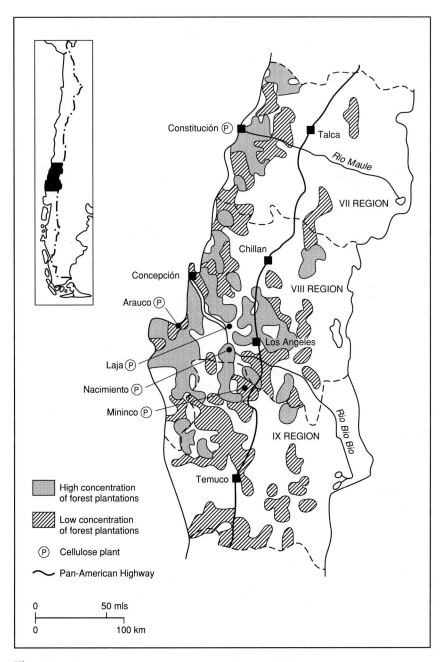

Fig. 9.5 Pine plantations and cellulose plants in Chile
Source: Gwynne (1993)

developed under ISI. Outward orientation has signified a major restructuring in economic sectors but, in general, primate cities have contained both declining inward-oriented and expanding export-oriented industries. In Chile, the shift to outward orientation has been associated with slight increases in the proportion of manufacturing employment in the primate city. These increases have not been in 'Fordist' factories but rather in small-scale, flexible enterprises. What Schmitz (1993) calls the collective efficiency of these small enterprises has meant that they have survived and adapted to the economic crises of the 1980s better than their larger counterparts. However, given the concentration of small firms in the primate city, the spatial distribution of manufacturing employment has not really changed, although its character has.

Further reading

Auty R M (1993) *Sustaining Development in Mineral Economies: The Resource Curse Thesis.* Routledge, London. (A detailed analysis of the problems (including Dutch disease) faced by mineral-rich Latin American countries in achieving sustained economic growth.)

Grosse R (1989) *Multinationals in Latin America.* Routledge, London. (A detailed and balanced analysis of the cyclical and often troubled relationships between multinational business and host country governments in Latin America.)

Gwynne R N (1985) *Industrialisation and Urbanisation in Latin America.* Croom Helm, London. (Examines the complex and varied links between the processes of import substitution industrialization and urbanization in Latin America.)

Gwynne R N (1990) *New Horizons? Third World Industrialization in an International Framework.* Longman, London and John Wiley, New York. (Compares the different processes of industrialization in East Asia and Latin America.)

Gwynne R N (1993) 'Non-traditional export growth and economic development: the Chilean forestry sector since 1974', *Bulletin of Latin American Research* **12**(2), 147–69. (Explores the sectoral and spatial development of an industrial sector able to compete in international markets.)

Gwynne R N (1994) 'Regional integration in Latin America: the revival of a concept?', in **Gibb R** and **Michalak W** (eds) *Continental Trading Blocs: The Growth of Regionalism in the World Economy.* John Wiley, Chichester. (Examines the continental context of integration in Latin America and the implications for industrialization.)

Hirschman A O (1968) 'The political economy of import-substituting industrialization in Latin American countries', *Quarterly Journal of Economics* **82**, 1–32. (The influential but flawed theory of industrialization through import substitution is synthesized.)

Hojman D E (1994) 'The political economy of recent conversions to market economics in Latin America', *Journal of Latin American Studies*

26(1), 191–219. (A useful and wide-ranging commentary on the political and economic context of the recent shift to outward orientation in Latin America.)

Katz J (ed.) (1987) *Technology Generation in Latin American Manufacturing Industries.* Macmillan, London. (A very revealing empirical survey of the prevailing technology of manufacturing under import substitution in Latin America.)

Kenney M and **Florida R** (1994) 'Japanese maquiladoras: production organisation and global commodity chains', *World Development* **22**(1), 27–44. (A useful case study of Japanese multinationals and their investment strategies in an outward-oriented Mexico.)

Lawson V (1992) 'Industrial subcontracting and employment forms in Latin America: a framework for contextual analysis', *Progress in Human Geography* **16**, 1–23. (An attempt to integrate small firm behaviour through subcontracting and employment patterns in Latin American cities.)

Messner D (1993) 'Shaping competitiveness in the Chilean wood-processing industry', *CEPAL Review* **49**, 117–37. (One of the few attempts to analyse the numerous problems that face national private enterprises in shifting from import substitution to outward orientation.)

Schmitz H (1993) 'Small firms and flexible specialization in developing countries', in **Spath B** (ed.) *Small Firms and Development in Latin America.* International Institute for Labour Studies, Geneva. (The most interesting attempt to link the small firm sector to the wider issues of economic and urban development in Latin America; theoretical perspectives backed by empirical research.)

Villaran F (1993) 'Small-scale industry efficiency groups in Peru', in **Spath B** (ed.) *Small Firms and Development in Latin America.* International Institute for Labour Studies, Geneva. (Develops a useful typology (influenced by De Soto) of small firms in Peru.)

Wong-Gonzalez P (1992) 'International integration and locational change in Mexico's car industry', in **Morris A** and **Lowder S** (eds) *Decentralization in Latin America.* Praeger, Westport. (A case study linking changes in the international economy and macroeconomic policy to locational change in a technologically dynamic industry.)

CHAPTER 10

Urban growth, employment and housing

Alan Gilbert

Urban growth in Latin America

Rapid urban growth is hardly a new phenomenon in Latin America. In Argentina, Chile, Uruguay and southern Brazil, several major cities had emerged by the end of the nineteenth century. Successful export production in these areas had led to the concentration of transport and commercial facilities in cities such as Buenos Aires, Montevideo, Rio de Janeiro and São Paulo. The pace of expansion was fuelled in each case by major flows of immigrants from Europe especially from Spain, Portugal and Italy (see Ch. 3). By 1920, Buenos Aires had 1.6 million inhabitants and Rio de Janeiro 1.2 million.

Elsewhere in Latin America urban growth had begun but was occurring much more slowly. Indeed, in the smaller Central American and Andean countries, where industrialization did not really begin until the Second World War, urbanization only started to accelerate during the 1930s. This rise was associated less with the pace of economic growth than with a general fall in the death rate. From the 1930s, the populations of most Latin American countries were increasing rapidly. At first, most of the natural increase was concentrated in the rural areas and was largely absorbed there, but, when population growth rates rose to over 3 per cent per annum, industrial development accelerated, opportunities for employment in the cities increased, and as land redistribution in favour of the poor failed to materialize, rural to urban migration got under way. By 1960, two out of five Latin Americans lived in cities with more than 20,000 people; in Argentina, Chile and Venezuela nearly two-thirds of the population fell into this category (Table 10.1).

The migrants were attracted to the cities by the availability of better paid work and encouraged to move by the increasing difficulty of making a living from small-scale farming. The opportunities available in the cities and the level of rural poverty obviously varied from region to region, but, even where there were major differences in rural and urban

Table 10.1 Urban population in Latin America, 1940–90 (per cent)

Country	1940	1960	1980	1990
Argentina	–	74	83	86
Bolivia	–	24	33	51
Brazil	31	46	64	75
Chile	52	68	81	86
Colombia	29	53	68	70
Cuba	46	55	65	75
Ecuador	–	36	44	56
Mexico	35	51	66	73
Peru	35	47	65	70
Venezuela	31	63	79	91
Latin America	**33**	**44**	**64**	**72**

Sources: Wilkie *et al.* (1991); UNDP (1992)

living standards, many people simply stayed where they were. This was especially true of the Indian populations of Ecuador, Peru and Central America, groups which would suffer most economic and social discrimination in the urban environment. Certainly, the population of the rural areas of northern South America continued to grow in absolute terms until the 1970s. The migration process was also highly selective within the rural areas. Young people who wished to further their education, new entrants into the job market and persons with skills such as bricklayers and drivers were more likely to move than older and unskilled workers. In most areas more women moved to the urban areas than men, largely because of the opportunities for unskilled labour available in domestic service and retail activities in the cities. Indeed, the differences in the numbers of men and women migrating to Latin American cities was often quite marked. In Bogotá, in 1973, there were 87 men to every 100 women and among the main migrant age group, 15–45 years, there were only 82 men.

Practically all urban centres gained from rural out-migration and after 1940 most cities grew faster than the rate of natural increase. Some small cities experienced dramatic expansions in their population. In Peru, Chimbote grew from a town of 5,000 people in 1940 to one of 173,000 in 1972, a result of a boom in fish-meal production and the establishment of a new steel works. In Venezuela, the population of the new industrial city of Ciudad Guayana increased from 3,000 people in 1950 to 320,000 in 1981. In Colombia, agricultural frontier towns such as Montería, Valledupar and Villavicencio quadrupled their populations in the 20 years after 1951, and along the Mexican–US border towns such as Tijuana, Mexicali and Ciudad Juárez saw their populations increase four or five times between 1950 and 1970. In terms of sheer numbers, however, the most spectacular increases were in the largest cities. By 1980, Latin America contained several of the world's largest cities; Mexico City had approximately 13.8 million inhabitants, São Paulo 12.2 million, Rio de Janeiro 8.6 million and Buenos Aires 9.7 million. Even previously small Andean cities had become major metropolises, Caracas had 2.6 million inhabitants, Bogotá 3.6 million and Lima 4.5 million. What is perhaps most startling was the pace of change. Between 1970 and 1980 Mexico City added almost 5 million people to its population, São Paulo 4.3 million.

Since 1980, the pace of expansion has slowed (Fig. 10.1). In particular the economic recession that hit most of Latin America in the 1980s had a major effect on the urban areas, particularly the largest cities. The import-substituting industrialization model which had strongly favoured industrial growth in the major cities was replaced with a model favouring export production. Manufacturing inefficiency in the major cities, long protected by trade restrictions and overvalued exchange rates, was suddenly punished by the inflow of cheap manufactured imports. Many companies went out of business, Mexico City, for example, lost around one-quarter of its manufacturing jobs between 1980 and 1988. Other privileges that had been garnered during the years of prosperity were withdrawn. Prices of food and transport rose rapidly as subsidies were removed. Employment became more difficult as factories and government offices laid off workers. Wages fell as inflation cut into pay packets and more companies went out of business.

As a result, the pace of city-ward migration slowed. It is estimated that net immigration fell from ten per thousand people in Santiago between 1977 and 1982 to two per thousand between 1987 and 1992. In Mexico City, CONAPO (1992) estimates that there was a net exodus of 300,000 people between 1985 and 1990. Clearly, this had a major impact on the rate of urban growth and for the first time in decades, the largest cities began to grow more slowly than many other urban areas, slower even than the national population (Table 10.2). Mexico City grew by only 0.9 per cent per annum during the 1980s, Rio by 1.0 per cent and Caracas by 1.1 per cent. Even the fastest growing city in Table 10.2, Lima, barely exceeded the rate of growth of the Peruvian population as a whole. In the process, the largest cities failed to reach the horrifying totals once forecast for them. Mexico City had only 15 million people in 1990, São Paulo 15.2 million in 1991.

Many among the new city dwellers were migrants, but, as the years have passed, much of the pace of urban expansion has been maintained by births in the cities. Since so many of the migrants were young they raised families in their new homes; no longer is rapid urban growth due mainly to migration. Indeed, the day has long been reached in many Latin American cities when migrants have ceased to make up the bulk of the population. This process of change is illustrated by the experience of Bogotá. In 1951 migrants made up 57 per cent of the population; by 1985, the proportion had fallen to 49 per cent. During the first half of the 1970s, the contribution of migration and natural increase to Bogotá's growth was approximately equal; during the first half of the 1980s, natural increase was twice as important as migration. This is reflected in the origin of migrants by age group: 76 per cent of people aged over 60 in 1985 had been born outside the city compared to only 46 per cent of those aged between 20 and 24.

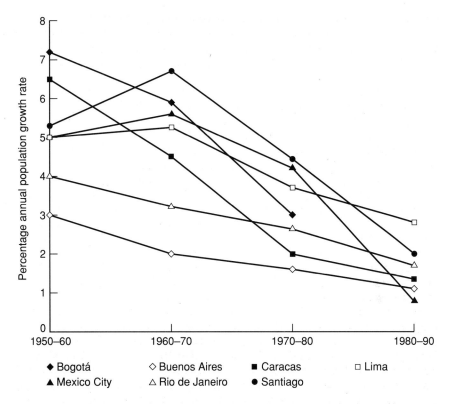

Fig. 10.1 Major Latin American city growth rates 1950–90 (*Source*: Villa and Rodríguez 1994: 54)

Table 10.2 Major Latin American cities: annual population growth rates, 1950–90

City	1950–60	1960–70	1970–80	1980–90
Bogotá	7.2	5.9	3.0	n.a.
Buenos Aires	2.9	2.0	1.6	1.1
Caracas	6.6	4.5	2.0	1.4
Lima	5.0	5.3	3.7	2.8
Mexico City	5.0	5.6	4.2	0.9
Rio de Janeiro	4.0	4.3	2.5	1.0
Santiago	4.0	3.2	2.6	1.7
São Paulo	5.3	6.7	4.4	2.0

Source: Villa and Rodríguez (1994: 54)

The major cities of Latin America are impressive not only for their size but because of the control they exert over the rest of their countries. By 1990, two out of five Argentinians lived in Buenos Aires, half of all

Uruguayans in Montevideo, one in four Chileans in Santiago and one-quarter of all Costa Ricans in San Jose. Any number of economic variables can also be used to show how the large cities of Latin America dominate their nations. In per capita terms there is more industry, more financial activity, more investment and higher incomes than in other cities. In 1987 Lima, with 28 per cent of Peru's people, produced 47 per cent of the gross national product and 69 per cent of the country's manufacturing output.

For a century or more, the level of urban primacy and the cities' economic dominance increased. It was not surprising insofar as the dominant model of development favoured them and many already had formidable advantages. Many were major ports (Rio de Janeiro, Lima-Callao and Guayaquil) and some also dominated the most productive agricultural regions of their countries (São Paulo, Santiago, Buenos Aires and Montevideo). Transport routes had spread outwards from these urban complexes giving them a major competitive advantage. In addition, since most higher-income groups lived in the major cities, these groups dominated the market for cars, refrigerators and clothing. Many of these cities were the seats of national government and benefited from the unprecedented expansion of the public sector during the period after 1945. Finally, as governments increasingly adopted an interventionist role in the development process, the numbers of workers in the civil service, in public utilities and in public transportation increased. As some trace of a welfare state emerged, more teachers, health workers and public administrators were employed. Not only did the location of government offer companies a market but it also eased the process of petitioning for import licences and price rises. The result was that, until 1980, commerce and industry became more and more concentrated in the major city regions.

Many governments expressed concern about the pattern of urban concentration. For a start, the simple arithmetic of population growth was a spectre to the governments of the continent; forecasts of Mexico City's population in the year 2000 regularly conjured with the figure of 30 million. Given existing levels of traffic congestion and air pollution, the lack of water and rising land values any further population expansion was of understandable concern. It is already very difficult to move about cities such as Caracas and the levels of air pollution in Mexico City and São Paulo would not be tolerated in most cities of the developed world. In addition, concern was frequently expressed about the difficulties that the expansion of these major cities put on the rest of their nations. Provincial politicians regularly complained that their regions were neglected by the concentration of investment in the major cities. They pointed out that the largest cities were growing most rapidly; while particular provincial cities and regions had grown spectacularly, most had lower growth rates than the major cities. Living conditions were also inferior in most smaller cities because levels of investment in public utilities are much lower. Similarly, there were constant complaints that the major cities were extracting wealth from the rest of the nation: provincial taxes exceeded central government spending in the regions; skilled manpower left for the large cities; centralism drained local autonomy and provincial dynamism.

However, if there were innumerable complaints about metropolitan growth, many advantages also derived from this pattern of urban agglomeration. Indeed, given the commitment of most Latin American governments to a combination of modernization and industrialization within a capitalist model of development, metropolitan expansion was almost inevitable. For many in the metropolitan centres there were few problems. Consequently, when governments attempted to decentralize manufacturing, for example, industrialists were rarely sympathetic. Even plans to move government workers away from the major city were problematic: the move to Brasília was hardly popular among Brazilian civil servants and, in Colombia, top administrators of one state corporation resigned rather than move to a provincial city. In Venezuela, the Guayana Corporation long had its head office in Caracas rather than in Ciudad Guayana.

As a result, numerous attempts to relocate industrial activity produced few notable successes. Many of these attempts were expensive and created few new jobs in the recipient regions. Even in north-east Brazil, the Venezuelan Guayana and the Mexican state of Guerrero, large new industrial complexes brought all too few benefits for the poor. Sometimes the results of deconcentration were almost laughable, for example, the incentives given to the Chilean car industry to develop in the far north of the country (see Ch. 9). More often than not, the policy simply failed.

Ironically, with the shift of the development model, much changed during the 1980s when many secondary cities had been growing more rapidly than the giants. There were two basic reasons for this change. First, with industrial development less dependent on the internal market and more oriented to export sales, the major cities had fewer advantages. Nowhere was this more dramatically demonstrated than in Mexico. While Mexico City's industrial base shrivelled as manufactured imports put many of its companies out of business, along the US border there was a manufacturing boom. Between 1976 and 1993, employment in the assembly plants along the border increased from 75,000 to 540,000. In addition, new export production facilities were established by transnational corporations in northern cities such as Chihuahua, Hermosillo and Saltillo. As Cordera and González (1991: 38) put it 'entire economic sectors have ground to a halt even as others – with connections to the external sector – thrive as never before. Areas with access to dollars from exports, the *maquiladora* industry, tourism, or migrants working abroad have been transformed into oases of prosperity in the middle of generalized economic stagnation.' The irony is that this process of industrial decentralization has occurred without government subsidies. The old, highly subsidized policy of industrial deconcentration failed, the new non-policy has produced the results that had been desired.

The second reason why the population of the major cities began to slow was that industry began to move out of the congestion and pollution affected central areas to small towns and cities in the nearby periphery. This process has been operating for some years around Buenos Aires, São Paulo and Mexico City, but the scale has grown recently and manufacturing plants have been moving much greater distances. As Storper (1991: 61–2) observes for Brazil: 'branch or assembly plants are undertaking a very different kind of decentralization: they are locating in industrial towns within a 200-km radius of the city of São Paulo such as São José dos Campos, Piraciciba, Americana, Limeira, Rio Claro, and Campinas.' This process has been encouraged by government policy only in the sense that increasing controls over manufacturing activity, for example, against air and water pollution, have hastened the move to smaller cities with less stringent controls.

Of course, this second form of urban deconcentration is hardly bringing benefits to most people in the poor regions of the country. For the poor regions to benefit there also needs to be a programme of rural development which embraces land reform and local participation in decision-making. As experience in the Brazilian north-east shows, merely enticing capital-intensive industry to poor regions does not help the rural or even the urban poor.

Employment

The most critical problem in Latin American cities is the lack of sufficient well-paid jobs. Were more such jobs available, the problems of housing, nutrition and health would be much reduced. Of course, many in the urban areas have regular and well-remunerated work. In addition to adequate salaries they also receive benefits such as health insurance, social security, retirement schemes and access to sports facilities. Most of these people work in the so-called 'formal sector' which includes the larger industries and offices, the bigger shops and commercial enterprises.

While there is no watertight definition of this sector, commonly used criteria include the following: corporate ownership, large scale of operation, capital-intensive technology, government-protected markets, formally acquired skills, dependence on foreign techniques and capital.

The relative size of the formal sector varies greatly from city to city. In the poorer countries, the cities contain relatively few large-scale industries and commercial activities; the professionalized middle-class and the blue-collar workforce are much smaller than in the region's major urban centres. It is difficult to be precise because of the difficulties involved in defining the formal sector, but in major cities such as São Paulo and Buenos Aires over half the labour force is involved in some kind of formal sector activity compared with less than a quarter in cities such as La Paz and Asunción.

Labour that is not employed in the formal sector is either unemployed or is engaged in some branch of the 'informal sector'. The level of open unemployment is often exaggerated in accounts of the situation in Latin American cities; urban unemployment is lower in most Latin American cities than is now characteristic of most developed world countries. It is true that at times of recession, levels of unemployment may rise to wholly unacceptable levels. For example, the level of open unemployment in Chile reached 20.1 per cent in 1982 and 14.3 per cent in Venezuela in 1984, but as Table 10.3 shows, this was atypical. As economic growth returned to the region, levels of open unemployment began to fall.

Table 10.3 Rates of open urban unemployment, 1970–92 (%)

Country	1970	1978	1980	1982	1984	1986	1988	1990	1992
Argentina	4.9	2.8	2.3	5.7	4.6	5.6	6.3	7.5	6.9
Brazil	6.5	6.8	6.2	7.7	7.1	3.6	3.8	4.3	5.9
Chile	4.1	13.3	11.7	20.1	18.5	13.1	10.2	6.5	5.0
Colombia	10.6	9.0	9.7	9.3	13.4	13.8	11.2	10.3	10.5
Mexico	7.0	6.9	4.5	3.7	5.7	4.3	3.5	2.9	3.2
Peru	6.9	8.0	7.1	6.6	8.9	5.4	7.9	8.3	9.4
Venezuela	7.8	5.1	6.6	8.2	14.3	12.1	7.9	10.5	8.0

Source: CEPAL (1982, 1993)

Of course, even when unemployment rates are relatively low in Latin America, this has little to do with the widespread availability of well-paid work. A more likely explanation is that most adults are obliged to find some kind of employment. An unemployed person cannot register for state benefits in the same way as in the United Kingdom, France or the USA. Quite simply, the choice for many is between finding some kind of work or going hungry.

Under these conditions, many of the unemployed have traditionally been drawn from the ranks of the poorest members of society. To be unemployed a worker requires family support; if a husband is out of work his spouse may work to provide the basic income. So long as an adult is working, a grown-up child can wait for suitable work; if a brother is employed, his sibling can remain at home. Many of those who are unemployed have technical or educational qualifications; for people with secondary school or university education it may be more sensible to continue searching for a suitable job than to take very low-paid work. If there is a reasonable expectation of obtaining a government job when the current administration hands over power, or of working in a private office because there is a regular turnover of people with the candidate's skills, then waiting will pay dividends. Indeed, for many professional groups in Latin America the irony is that there are more trained workers than jobs. In Colombia, there is a crying need for doctors and yet, because of financial constraints, many trained medical practitioners cannot obtain jobs; they either seek work in other activities, such as business or teaching, or wait for friends with political influence to obtain a post for them.

This situation was much more true of the 1970s than of the 1980s.

253

The deep recession that affected most of the region's urban areas so badly during the 'lost decade' forced large numbers of workers out of work. However, even though there were more unemployed, the greatest impact was to enlarge the ranks of the so-called 'working poor'. The recession increased the numbers of low-paid workers. Many families survived only because they put more workers into the labour market. In Mexico, González de la Rocha (1990: 118) argues that 'to avoid a drastic reduction in food consumption, households have sent more members to the job market; more youths, women, and children have entered the work force in order to earn the income needed for the survival of the group, the domestic unit.' Hakkert and Goza (1989: 91) state that 'In some more desperate areas, even the participation rates of children increased in an effort to help their families reach a minimum level of subsistence'.

The recession also increased the number of working poor in another way; by directly reducing their incomes. As Table 10.4 shows, the real manufacturing wages in cities fell dramatically during the recession. In Lima, manufacturing workers in 1990 were earning only 43 per cent of what they had earned ten years earlier. They were not underemployed, because most work very long hours for little reward, simply engaged in low productivity work. Fewer people could do the same amount of work, but the paradox is that they are still required to work a long day. Take, for example, the case of numerous street vendors who sell sweets, fruit and cigarettes. Insofar as custom comes regularly through the day, long hours are required; cut the working day and the worker's income will fall. Hence, although sales per worker are limited, the working day is long. Underemployment is obviously not the best term to describe the hours worked by most Latin Americans and the more emotive term 'working poor' is perhaps more appropriate.

Table 10.4 Changes in real incomes, 1978–92 (1980 = 100)

Country	1978	1980	1982	1984	1986	1988	1990	1992
Argentina	78	100	80	117	102	93	80	76
Brazil	97	100	116	105	122	103	88	106
Chile	85	100	109	97	95	101	105	115
Colombia	93	100	105	118	120	118	113	117
Mexico	106	100	108	75	72	72	78	85
Peru	95	100	93	87	98	76	36	43
Uruguay	109	100	107	72	72	76	71	75

Source: CEPAL (1985, 1992b)

The informal sector

Many of the new working poor have been incorporated into the so-called 'informal sector'. Drawing on the earlier experience of severe recession in Chile, Lautier (1990: 289) shows how the 'survival informal sector' expanded to sustain the poor: 'from 1971 to 1982, the proportion of the industrial workers in the labour force fell from 21 per cent to 11 per cent. Informal employment rose from 18 per cent (1970) to 27 per cent (1982).'

While Lautier's interpretation of the direction of the trend is undoubtedly correct, the precision of his figures is undermined by the fact that we are anything but clear what activities should be included under the term 'informal employment'. Originally, the term 'informal sector' was devised to emphasize the significance of self-employment and small-scale enterprise in providing work in Third World cities. It was subsequently adopted by the International Labour Office which sought to assist the informal sector in an effort to increase both productivity and the number of jobs available. The basic problem with this concept is that, like most dichotomies, the distinction between the formal and the informal sector is difficult to sustain. While iron and steel plants are clearly part of the formal sector and bootblacks part of the informal sector, most activities are less easily classified. The informal sector is especially problematic because it is a residual category for everything excluded from the formal sector. As such it contains a rag-bag of activities with little in common. This imprecision has always impeded policy formulation.

Early interpretations of low-wage work had suggested that most jobs were 'involutionary'; the jobs were unproductive because they merely subdivided existing tasks so as to produce work for the unemployed. New jobs were created not because of new needs but because existing functions were subdivided. This categorization of work suggested that informal sector jobs were small-scale, labour-intensive, lacking formally acquired skills and reliant on domestic resources. Most critically, jobs were easy to enter and markets were unregulated by governments. In fact, detailed studies of Latin American cities have revealed the diverse array of jobs contained within the informal sector. Many of these jobs fit uneasily into the classification because they can be entered only with difficulty and often require experience, capital and permits. Indeed, the informal sector appears to be highly structured with a well-developed hierarchy of functions and seniority. Thus salesmen in the city centre often require licences which have to be bought from the authorities or from former salesmen. Prime sales positions are reserved for those who have been longest in the profession or who can afford to buy the rights to the spot. Bootblacks cannot practise in the best locations without competition with existing workers; cleaning the shoes of clerical staff in an office is the prerogative of a regular bootblack – the porters will not allow a new bootblack into the building. Clearly, the better jobs in the informal sector require experience and contacts. It is no more possible to enter the most remunerative parts of the informal sector than it is to start off as managing director of a formal sector company.

In addition, most of the better paid informal sector activities require capital. Selling lottery tickets requires that a salesman has the money to pay for the tickets in the morning; he is not repaid until he has sold them. Many informal sector workers avoid this constraint by working for someone else, for borrowing money is very expensive. Clearly, ease

255

of entry into better paid parts of the informal sector cannot be achieved without difficulty.

The informal sector does not consist wholly of so-called 'petty services', bootblacks, salesmen, newspaper vendors, beggars, etc. It also comprises the artisan and craft workers, many small garages and repair shops, and other clearly manufacturing activities. Given the wide-spread perception that services are 'unnecessary' and manufacturing 'necessary', the discovery that the informal sector also contains the latter helps convince sceptical government officials that the sector contributes to the urban economy. Indeed, more and more literature is contradicting the original idea that the informal sector is in some sense parasitic on the urban economy.

Early work on Latin American cities constantly suggested that jobs could be eliminated without any loss in production. These activities were marginal to the main sources of employment and production, and existed only so that people could survive. In addition, it was thought that there was little interaction between the 'productive' and the 'unproductive' sectors. These ideas encouraged governments to provide finance and training for the productive formal sector and to ignore or even discourage the informal sector. The consequences were that many formal sector activities were given easy access to public and private financial agencies, whereas small informal sector activities were neglected; many street traders and small businesses were constantly harassed by the police because they lacked licences. So long as the image of the informal sector was one of social rather than economic utility, such policies were likely to endure. Increasingly, however, this assumption came under scrutiny, and during the 1970s the World Employment Programme came to regard the 'informal' sector as a source of opportunities for creating more adequately paid jobs. Advice, credit and new techniques were fed into different kinds of informal activities as a method of creating more labour-intensive jobs. Governments were convinced that harassment of the sector was less appropriate than encouragement.

Other views of the informal sector are also emerging which question additional aspects of our understanding. The belief that the formal and informal sectors operate independently has been a major casualty. Indeed, several writers have argued that profits in the formal sector depend upon the existence of the informal sector. One study in Cali, Colombia, showed how the paper, bottles and waste metal that are collected from the municipal rubbish dump are recycled through the big industrial companies in the city (Birkbeck 1978). One of the most 'marginal' of activities is shown to be linked to the formal sector, and, because the recycled materials are cheaper than producing new paper, glass or metal, some benefit is clearly derived by the industrial companies from the transaction. Manufacturing companies also benefit in terms of sales from the informal sector. Cigarettes and sweets are sold on most street corners, indeed the sales distribution system could hardly be improved, and yet the companies do not formally employ their sales 'staff'; they do not pay their social security, they pay only for results and laying off staff is no problem. The result for the companies is a cheap, flexible salesforce; for the poor, long hours and low wages.

The argument can be taken further to argue generally that the cost of labour in the formal sector is reduced by the presence of the informal activities. First, costs of living of factory workers are kept down by the low price charged by shopkeepers and marketsellers operating in the low-income areas of the city. Second, the self-help housing, in which many formal sector employees live, provides cheap accommodation which keeps labour costs down. In this sense, the formal manufacturing sector may be gaining throughout the production and sales process – it obtains higher profits because of cheap inputs and because it obtains a cheap, flexible sales system. In addition, it often persuades governments to act against the informal sector when the latter's activities have ceased to be useful to the formal sector.

Of course, it is erroneous to take this interpretation too far. Low incomes in the informal sector reduce the costs of production and sales of the formal sector, but they also mean that its market is much more limited than would be true if there were a highly paid labour force. How the balance of advantage works out for an individual company is difficult to evaluate; some producers clearly benefit, others lose. For example, companies producing manufactures for export gain twice over; they have a cheap labour force and their market is unconstrained by low domestic wages. By contrast, the domestic market for manufacturers of clothing or refrigerators is limited because of the low incomes of the urban poor.

Whatever the balance of advantage the critical point is that the formal and informal sectors are intimately linked in the Latin American urban economy. The poor are not marginal to the 'productive' sector but form part of it. Their jobs may not form part of the formal sector, but their work adds to its profitability. Nor are most of the jobs the traditional activities of former peasants. While some craft production persists, most informal activities are modern jobs. Selling American brands of cigarettes, repairing cars, helping shoppers carry their purchases from the supermarkets to their cars and installing electricity or plumbing into self-help homes are not activities that could have existed in 1900. These jobs are not only modern, they are also economically productive and far from 'marginal'.

Social segregation in the city

If many Latin American families are poor, there is also a minority of rich ones. Most Latin American cities exhibit very wide differences in household incomes. The skills of many professional groups in Latin America guarantee them a good income. If in the formal sector there is no shortage of competition, the relative shortage of engineers, accountants, computer programmers and surveyors guarantees them rewards that are often higher in purchasing terms than those received by comparable groups in developed countries. In addition, a minority earn large incomes through their ownership of stocks, land and real estate. Thus the distribution of income in urban areas is very unequal, the majority surviving on low incomes with a minority gaining major advantages from their market position or their control of capital. During the 1980s, levels of inequality got worse. In Buenos Aires, the average income of the richest decile was ten times higher than that of the poorest decile in 1984. By 1989, the difference had risen to a startling 23 times (Ainstein 1994).

One result of such marked differences in income is that all Latin American cities have clearly segregated areas of land use. There are industrial zones which accommodate modern factories, well-developed commercial and rental centres, high-income residential areas, zones of government and private offices, and large swathes of low-income residential development. Some parts of the cities are well ordered and regulated, others lack services and appear to have developed quite spontaneously. This segregation is mainly the result of market forces. Industrial companies and high-income residential groups can afford to bid for high-value land, the poor occupy residual land: the market mechanism is modified by the intervention of governments although most public agencies are forced to buy land like any private company. Governments do influence the price of land through their servicing and planning policies and through taxation. Sometimes, too, governments control large areas of land, a legacy sometimes of colonial rule, as in Venezuela's municipal land, of revolution, as in Mexico's *ejidos*, or of the fall of a dictator, as in Nicaragua or Cuba. In general, however, the dynamics of urban land use are determined by the interaction of demand and supply, and government intervention is rarely effective.

Most Latin American cities have a central core which developed during the colonial period sometimes on the site of an existing settlement, as in Mexico City and Cuzco, but most frequently in a new location. The centre piece of the Spanish American city was the colonial *plaza mayor*, a large square flanked by the cathedral or church, government offices, and other public buildings. The homes of the élite were originally located close to the square along the streets which spread out in a grid-iron pattern. The lower income areas were further out at the edge of the city. Portuguese colonial settlements usually lacked a central plaza, were often less regular in design, and were more likely to follow the dictates of the terrain, even if the differences with the Spanish settlements have often been exaggerated.

Until quite recently, most cities maintained this urban form. It has changed only as a result of changes in housing design, new forms of transportation, the growth of car ownership, and the rapid expansion of the urban population. During the twentieth century, élite groups have gradually moved away from the central city to occupy suburban housing. The growth of large factories has required the reservation of special industrial areas close to railway tracks or main roads. As cars and also buses have come to dominate urban travel patterns, new commercial areas have developed along the main routeways between the city centre and major residential areas. Governments have responded by improving road communications, attempting to separate conflicting land uses, and generally regulating the pattern of urban expansion. While many cities still exhibit high population densities by North American standards, there has been a strong movement outwards towards a more dispersed suburban pattern of design, with segregated activity areas.

The dynamics of urban growth are determined by commercial demand and by the market power of high-income residential groups, although some land is reserved for necessary government functions and public services. The land that remains is available for use by those groups that can afford to pay least. Hence the urban poor occupy land that is unpopular among other groups. Such land is liable to flood, suffers from some kind of pollution, or because of its physical characteristics is difficult to service. Whatever the mechanism by which the poor obtain land, they occupy the areas that other groups have left unoccupied.

Planning and servicing policy tends to accentuate the patterns of segregation that result. The service agencies first supply areas where the owners are politically powerful and/or can afford to pay the cost of the services; industrial areas and higher-income residential areas are always fully serviced. The location of service lines and roads affects the price of land and helps to determine neighbouring land uses. New high-income residential areas will develop close to existing élite areas both to gain access to services and to share the social cachet of the prestige locations. Thus, land values and the way that the land is divided by developers into large or small plots will determine the future patterns of land use. In low-income areas, the lack of adequate services will discourage most higher-income groups from moving in. The combination of market forces and servicing policy accentuates social segregation. The level of segregation can be demonstrated by the maps of Caracas and Bogotá. Both show how high- and low-income residential areas occupy different sectors of the city. In Caracas, the low-income *rancho* areas congregate on the hillsides, occupying land that has been avoided by other land uses. In Bogotá, the low-income areas are concentrated in the south of the city and to the north-west. The environmentally more desirable areas to the north are the domain of high-income groups; one poor area located in the north is contiguous to a limestone quarry and cement works. In the south, the land is either liable to flood or is located on hillsides that are difficult and expensive to service.

The housing of the poor

Two stereotypes of poor housing appear in the literature: the one relating to conditions in the central areas of large cities, the other to the peripheral shanty towns. The first picture derives from detailed anthropological observations of conditions in the central cities of Latin America where newly arrived migrants and recent-established native households rent accommodation in crowded multifamily occupancy. Many of the houses were designed for high-income families which have since moved to more modern, suburban accommodation, their former residences having been divided into self-contained rooms to rent. Elsewhere 'purpose-built' accommodation has been constructed for rent. These *conventillo* or *inquilinato* areas contain high population densities and living conditions are poor. One such area in Mexico City, with 700 inhabitants living in a single block, is described by Oscar Lewis (1963: xiv):

> The Casa Grande is a little world of its own, enclosed by high cement walls on the north and south and by rows of shops on the other two sides ... This section of the city was once home of the underworld, and even today people fear to walk in it late at night. But most of the criminal element has moved away and the majority of residents are poor tradesmen, artisans and workers. Two narrow, inconspicuous entrances, each with a high gate ... lead into the vecindad on the east and west sides ... Within the vecindad stretch four long, concrete-paved patios or courtyards, about fifteen feet wide [5 m]. Opening on to the courtyards at regular intervals of about twelve feet [4 m], are 157 one-room windowless apartments, each with a barn-red door ....
>
> In the daytime, the courtyards are crowded with people and animals, dogs, turkeys, chickens, and a few pigs. Children play here because it is safer than the streets. Women queue up for water or shout to each other as they hang up their clothes, and street vendors come in to sell their wares.

The second stereotype is of the owner-occupied shack perched on a hillside or precariously balanced on wooden piles which extend outwards into a river, sea or lake. The house is built of flimsy materials and has been built on land which is unstable or liable to flood. The services have been obtained through illegal hook-ups to the main electricity or water supplies. The only conceivable sign of prosperity is the sprouting of television aerials from the top of the houses.

Both pictures are accurate in the sense that too many Latin Americans live in primitive conditions. They are also accurate in the sense that far too many inhabit homes that are totally unsatisfactory by European standards. At the same time both pictures are misleading of the general conditions in which most poor urban dwellers actually live, a point illustrated in the following sections.

Self-help housing

The term 'self-help housing' is not easy to define precisely but refers to the large areas of Latin American cities where the dwellings were built by the original occupiers on land suffering from some degree of illegality and where no services were initially provided. The term is used in contrast to the legal, serviced, architect-designed and company-built houses of the 'formal sector'. The proportion of the population living in self-help housing varies according to the income of the city, the pace of urban expansion, the policy of the state towards housing tenure and illegality, and finally the ease of access to land. In high-income cities, few families choose to build their own homes; in very low-income cities few can afford the costs of construction materials. In slowly expanding cities, there may be enough conventionally built houses to accommodate most families; in rapidly expanding cities there is little alternative to self-help construction. Even where the lack of housing invites self-help, however, the state may discourage the process. The critical element here is the state's attitude to land. The popular idea is that the urban poor occupy land through large-scale invasions despite the opposition of the authorities. The reality is, however, that invasions typically occur only where they have the covert support of politicians or government administrators. The corollary, of course, is that in some cities invasions are not permitted.

In those cities where invasions are actively discouraged, the poor are forced to purchase land. Of course, they cannot afford to buy land in good neighbourhoods and can rarely buy land that is serviced. The only real alternative is to buy land that is illegal in the sense that it has not been sanctioned for urban use by the planning authorities. Such land has not been developed because it lies beyond the urban limits of the city, because it is unserviced, or because the land cannot legally be sold for urban use. The forms of illegal land purchase vary from city to city. On the edge of Mexico City *ejidos* are sold illegally by members of the community to the urban poor. In Bogotá, sales are organized by an illegal subdivider who sells plots of land on the fringe of the city. The formation of the illegal subdivision of Britalia in Bogotá illustrates certain of the features characteristic of the low-income residential land market.

Britalia is located in the south-west of Bogotá on low-lying land which is liable to flooding in winter. It is flanked to the west by open land belonging to a large estate and to the east by other illegal subdivisions. Sales began in 1973 but possession was not granted until 1974 when most lots were in fact sold. The pirate urbanizer bought the land, or at least promised to buy the land from the two owners. By 1976 he had paid off the smaller debt but still owed the bulk of the purchase price. The urbanizer hoped that the settlement would be legalized under the 1972 Minimum Standards Decree but this was not permitted because one-fifth of the planned area lay outside the urban perimeter and because the water company felt it would be difficult to drain the lower parts of the settlement. The request was formally refused in January 1975. In May of that year the Committee for Community Action denounced the urbanizer to the authorities. The latter did not intervene, officially because they lacked confidence in the state authority (ICT) to service the settlement, and signed an agreement with the urbanizer to legalize the community providing he brought the settlement up to minimum standards servicing levels. He complied with part of the request, paid for the installation of electricity, paid off part of his debt to the original owners, and had made considerable efforts to install some water and drainage services. His efforts were not sufficient, however, and in June 1976, urged on by the community, the authorities took over the financial running of the settlement. By 1979 the settlement had around 1,600 families who occupied half of the 2,846 lots available. The mean purchase date of the original households was towards the end of 1974 and the mean moving-in date the beginning of 1976.

In other parts of Latin America, of course, invasions are permitted by the authorities in certain parts of the city. Sometimes the invaders have been covertly organized by powerful personages in the state. In Peru, General Odría, president from 1948 to 1956, encouraged the poor to invade land in Lima as a means of undermining the support traditionally received by the popular APRA party. Sometimes prior government authorization allowed the land to be occupied during the day with government vehicles carrying the families to the site. Sometimes the link was more covert and the invasion occurred at night. On other occasions opposition political parties organized the invasion which the government was later forced to accept to maintain its public image.

In Venezuela, the state is also heavily involved in the invasion of land as a means of obtaining popular support. Indeed, ever since a dictator in the 1950s destroyed large numbers of *ranchos*, the political parties of the country have encouraged land invasions. While the government may discourage invasions during much of its term of office, under the stress of elections invasions proliferate.

Even in countries where invasions are permitted, many efforts to occupy land are repressed. Such repression may occur for a variety of reasons. Sometimes the organizers of the invasion have failed to acquire adequate political support. It is very common, for example, for minor politicians to try to organize invasions to build up a personal following. If they can convince a large group of people to occupy land and the police can be discouraged from removing them, the politicians can count on the backing of the population, especially when they are able to persuade the authorities to provide services. Minor politicians may face opposition from more powerful colleagues, or from members of other political parties. If the police are controlled by the opposition, then the invaders may be removed. If the land that is occupied is a source of embarrassment to powerful political allies, the result may be the same. The best recipe is to choose the site carefully: land that belongs to foreigners, especially at a time of national xenophobia, land which is subject to doubts about its ownership, which the owner fails to guard adequately, or which has little commercial value. It is not advisable to occupy land that is contiguous to high-income residential areas or to invade farms belonging to powerful political families.

The police have been sent in to remove innumerable invasion settlements in Latin American cities. The outcome of their intervention has depended upon the factors just mentioned but also on the nature of the subsequent confrontation. Not infrequently death or injury to some of the invaders has been sufficient to persuade the authorities to recall the police. Where the authorities have been determined, of course, even such a tragedy has had no effect on the outcome. In Caracas, during the 1950s President Pérez Jiménez not only opposed new invasions but demolished large areas of existing shanty housing. The National Guard was used to drive the occupants out of the *ranchos* into newly constructed apartment blocks, a wholly misguided attempt to improve living conditions for the poor. More recently, the authorities in Chile

Plate 10.1 Small houses beside tower blocks. Cheap housing, once informal, now at least semi-permanent, huddles beside smarter apartment blocks on the northern margin of Buenos Aires, 1995. (Photo: D Preston)

vigorously opposed land invasions. In Santiago, when a series of attempts to occupy land occurred between 1980 and 1985, all but 3 out of 24 invasions were removed. In 1984, a violent conflict between the police and invaders in Puente Alto led to the death of 2 people and injury to a further 32.

Once land has been occupied, whether through invasion or through illegal subdivision, the process of settlement development is very similar in most cities. The initial problem is to construct a shelter. Where people have invaded land it is essential to build quickly and to live on the site during construction; in illegal subdivisions construction can proceed at a less frenetic pace. Materials are piled up on the sites and different households build a one or two room house as fast as possible. At first the shelter may be quite flimsy but within a couple of years solid wood or brickbuilt structures will appear. In cities with hot climates, straw or canvas will be replaced by wood or corrugated iron; in highland cities or countries with cold winters brick will be more common. Most families will engage in construction themselves but the extent to which houses are self-constructed will depend on the income of the family, their skills and experience, and the time at their disposal. Few families fail to enlist the help of skilled friends and neighbours for

263

special tasks such as putting in glass, bricklaying, installing electricity points and plumbing; most will pay for specialized help.

Construction of the house is a matter for the individual household, but all families have a common interest in settlement consolidation. Many are likely to belong to, or be involved in, a community organization which petitions for services and organizes communal projects. Electricity and water are the first priorities and the community will approach the authorities and friendly politicians to arrange matters. Where help is not forthcoming or where a long wait is expected, the community may establish illegal electricity lines. Skilled workers will be contracted to acquire the transformers and to link the settlement to the mains. Each family will then pay to be connected to the transformer.

The community and the authorities will negotiate for several years over the provision, legality and costs of services. Large settlements are likely to find favour more quickly, especially when elections are close. In Venezuela strange changes of policy occur close to elections; earlier decrees about removal are reversed, water is supplied to previously unfavoured settlements. Gradually, the neighbourhood receives more and more services and after ten years there is little that distinguishes the self-help settlement from the housing found in the lower-income areas in the rest of the city.

The pace of consolidation of course depends upon the income of the families relative to the costs of materials and land. Where materials are expensive and land has to be purchased, the pace of consolidation may be slow. Sometimes it can be very rapid, especially when middle-income families move into the settlement. Once services have been installed and the houses have become consolidated, families will begin to rent out rooms to other families. Frequently, rooms with a separate front door will be constructed. Gradually, the proportion of tenants in the settlements increases as more families rent out accommodation and increase the numbers of rooms they have available to let. Indeed, this process is now so widespread in cities such as Bogotá, that most of the tenant population live not in conventionally built housing but in consolidated self-help areas. As the population of the city has increased, the old central areas have been unable to accommodate the majority of the new households especially as some of these areas have been demolished to make way for new office and commercial developments.

The tenant population

According to the well-known housing model developed by John F. C. Turner, the tenant population consists mainly of recent migrants to the city who establish themselves in the home of kin or rent accommodation. Once they become established in the city with regular work and a growing family they contemplate moving into the urban periphery where they can engage in home construction. Over the years this has led to a dramatic increase both in the proportion of the population living in self-help housing and in the level of owner-occupation in Latin American cities (Fig. 10.2). By the early 1980s, most

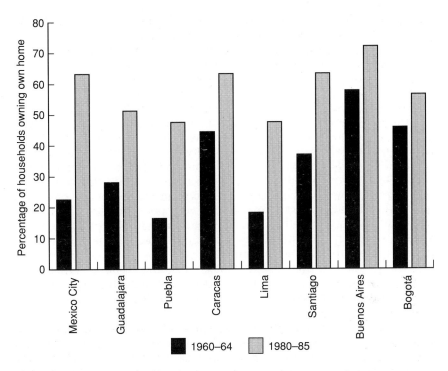

Fig. 10.2 Per cent of households owning own home (*Source*: Gilbert 1994)

households in Latin American cities owned, formally or informally, their own home.

However, despite this dramatic shift, it has been suggested above that the move into owner-occupation is not always a matter of choice. Potential self-help settlers need to have accumulated sufficient savings to pay for the cost of land and materials. Of course, where land can be invaded easily the total cost of home construction and consolidation is relatively low. However, in cities, such as Bogotá, Medellín and São Paulo, where land cannot be invaded and even plots in illegal sub-divisions are expensive, many people have to wait a number of years before they can afford to buy land. Indeed, in Bogotá in the late 1970s, the average age at which home consolidators first obtained a plot of land was 36 years. This relatively advanced age suggests that access to self-help housing is not easy and that many households are forced by their lack of savings to live many years as tenants.

However, if some tenants are forced to wait before they move into owner-occupation, others choose to wait. Indeed, in some cities the poorest households are often the most recently established self-help owners. The picture of impoverished tenants and socially mobile self-help owner-occupiers conjured up by the Turner model is sometimes misleading. Surveys in several Latin American cities now suggest that among the urban poor, tenants are rarely worse off than owners. In low-income settlements in Caracas, Mexico City and Santiago, the *household* incomes of owners and tenants were very similar but the smaller size of tenant households meant that they had higher *per capita* incomes. New owners were particularly poor when compared with long-established tenants or even when compared with households sharing with kin.

How is it possible that although most households declare that they wish to become home owners, tenants with higher incomes remain in rented rooms when many poorer families move into self-help accommodation? The most likely explanation is that the more affluent tenants can afford to wait until they can find the kind of property that they want to live in. This may be a semi-constructed house or a plot which already has services. By contrast, very poor families who cannot afford to pay their monthly rent may be forced to occupy land in the distant periphery; a particularly likely outcome, of course, in cities where land occupations are permitted.

Some tenants are prepared to wait because the accommodation they occupy is superior to what they would have in the self-help periphery. Most rental accommodation has water and electricity, it may be small but it is solidly built. Nor is their tenure desperately insecure. In some cities, such as Mexico City, rental accommodation is even rather cheap. Frequently, rental tenure is also quite secure; while evictions are not uncommon, many tenants have lived in the same accommodation for years. In the centre of Mexico City, 73 per cent of tenant families have lived in the same accommodation for more than ten years; in the centre of Santiago, 46 per cent. Many landlords and tenants even speak well of one another.

Of course, tenants do face problems, and many do want to become owners. In cities such as Santiago it is very expensive to rent. In places, landlords do operate in high-handed fashion and exploit tenants. Most seriously, however, is the fact that few governments have policies designed to improve and extend the rental sector. Many former tenants have moved into owner-occupation because their landlords have given up renting. Only gradually are governments beginning to realize that if something is not done to encourage investment in rental housing in Latin American cities, a key component of the housing stock will be in danger. Arguably, every city needs rental housing; without it, families have too limited a housing tenure choice. Not everyone wants to own; many more are not in a position to do so.

Urban planning

Urban planners were active in the Americas well before the arrival of Columbus. The great Aztec and Inca civilizations produced ceremonial cities, such as Cuzco and Teotihuacán, planned meticulously by architects and artists under orders from an élite. With the coming of the Spanish and Portuguese, planning changed in nature but continued to dictate the form of the rapidly changing built environment. The hundreds of new cities that were built by the Spanish and Portuguese mostly followed plans based upon the urban designs that had matured in the mother countries. Most urban plans were very rigid and strictly implemented; they laid down the precise shape of the city, its houses and streets; they determined which areas were to be occupied by different social groups.

In recent years planning has been practised often not because it would create a better human environment but because it could be used as an additional weapon in the armoury of the powerful. Let us consider, for example, the effect that urban planning has had on social segregation. Much of the literature on urban planning has argued that different social groups should live within the same neighbourhoods. Many urban planners have acted as social engineers striving to reduce inequalities in society. Within urban Latin America, however, practice has been different; there has been a 'deepening of class divisions within these cities, reflected in increasingly well-defined patterns of spatial segregation by social class' (Cornelius and Kemper 1977). Of course, such segregation might have been more acute without the intervention of planners. It is possible that residential differentiation was inevitable given such wide differences in per capita wealth and income and that urban planning has simply made little difference, but there are also depressing examples where planning may actually have accentuated the level of segregation. As Cornelius (1977) notes: 'most government interventions (public housing projects, urban renewal schemes, zoning regulations, investments in urban infrastructure, and provision of public services) have only reinforced the market forces promoting class segregation ... In fact, most governments seem intent upon "building the divided city".'

Specific cases where government intervention has accentuated residential segregation are easy to find. In Bogotá, practically all government housing schemes have encouraged the natural tendency towards class segregation. Urban land-use zoning may have had a similar effect. Indeed, Amato (1968) argues that such zoning was introduced into the city to maintain segregation. It was introduced not with reformist zeal but as a response to alarm among the élite that 'their new residential areas might be undermined by the encroachment of undesirable uses' at a time of rapid urban growth. He claims that it achieved its purpose: 'the early zoning of the city permanently fixed the spatial distribution of socio-economic groups and bears testimony to the desire and ability of the élites to structure the city in conformity to their own interests'. If this view overstresses the influence of élites in the formulation of the zoning scheme it shows how zoning helped these groups insofar as it protected their residential areas.

In Chile, urban planning under the Pinochet regime was much more blatant in the way that it increased social segregation. Not only did the Pinochet administration oppose new land invasions but it displaced many existing settlements. Between 1979 and 1985, somewhere around 30,000 families were moved from land belonging to private owners. The eradication programme led to a major shift in the location of the *campamento* population. Many were removed from the affluent north-east to the poor south and north- west of the city. The relocation programme created more homogeneous communities, thereby accentuating the already extreme level of residential segregation in the city.

Such negative examples merely show that urban planning is capable of worsening the situation. Unfortunately, it is seldom able to improve living conditions for the poor because alone it is incapable of modifying the social structure or the power system. If sensible urban planning were introduced into an equitable society in which all social groups participated in decision-making, its effects on the urban fabric would be very different. The reality in most of Latin America, however, is that élitist regimes rule in the interest of limited groups of the populace. In such circumstances it is perhaps not surprising that planning has often been misused. For example, urban planning has often been adopted only as a means of obtaining foreign loans, a practice tied to the trend in recent years for agencies such as the World Bank to insist that loans be linked to rational and technocratic forms of decision-making. As such, some governments have embraced urban planning not as a means of restructuring urban society but as a mechanism for obtaining hard currency credit; planners have been used merely to dress government projects in suitable clothes to parade before the international lending institutions.

Of course, numerous efforts have been made to use urban planning in the redesign of urban society. Unfortunately, few of these efforts have got very far, for they have suffered from a variety of difficulties. First, many cities have grown beyond their original administrative boundary and now occupy several political areas. Such administrative fragmentation hinders effective land use control, public service delivery and most kinds of forward planning. Second, Latin American cities seldom have sufficient resources to finance the costs of services and infrastructure. Attempts to raise taxes are often blocked by the rich and powerful and, without an adequate tax base, essential services cannot be provided for the poor; health, education and social welfare facilities are unsatisfactory for most low-income groups. Third, the very nature of government administration in Latin America undermines many efforts to plan urban development sensibly. High-income residential developments often ignore the planning regulations, usually without sanction. Bribes or influence deflect the attention of the regulatory agency or convince the planning authority to change the regulations: low-income settlements may 'splice' into the main water pipes and be protected by a political patron. In Latin America, the line between partisan politics and administration is much more blurred than in the United Kingdom.

In short, urban planning fails to resolve the problems of urban society. It fails both because urban problems are complex and pervasive, and because planning mechanisms reflect the balance of power in Latin American cities. Since the state is rarely truly democratic, state agencies do not often act in the interests of the whole society. Sometimes, government officials directly help the poor, but such interventions are rare. Sometimes, the poor gain indirectly from plans intended to help other groups; expansions to the water supply demanded by industry may bring improvements to domestic water provision. But, generally, urban planning is merely one more mechanism by which the rich and powerful manipulate decisions to their own benefit.

Clearly, better planning is required in the future if some current urban problems are not to get worse. There needs to be better implementation; planning regulations are ignored too frequently both by the private sector and by public agencies. In Latin American cities there always seem to be good reasons why absurdities should be allowed to continue. In Bogotá, private transport companies still refuse to introduce bus stops along most of their routes because they maintain it would cut their takings. The effect on traffic flows and accident rates as buses swerve across lanes to the kerbside to pick up additional passengers can be imagined. Throughout Latin America, urban land use regulations are ignored in the poorer parts of most cities because if they were enforced, most self-help housing would have to be demolished. Clearly, sensible planning for transportation, housing and servicing becomes ever more important as urban populations increase. While there have always been many instances of adequate planning, it has never been a consistent trend. Many cities which have requested loans from the major international lending institutions have been forced to modify their planning processes. Most of the major electricity and water improvement projects that have occurred in Latin America's larger cities have been financed from abroad and loans have often been subject to strict regulations about pricing, organization and efficiency within the public utilities. Whether, of course, pressure of this kind is a wholly positive development is a fascinating area for debate but change is certainly occurring to rationalize the provision of these services.

The major solutions to the problems of urban expansion, however, lie outside the realm of urban planning. They lie in changes to the model of development that has been adopted in most Latin American countries; many of the urban problems are a direct consequence of national policy decisions. Acute traffic congestion, for example, is a corollary of the decision to maximize domestic car production by encouraging home consumption. Poor housing conditions are part of the price that must be paid for the refusal to redistribute income and to claw back rapid rises in the price of urban land. The very process of urban growth is an outcome of the failure to redistribute agricultural land to the peasantry, the decision to favour urban activities in investment decisions, and the preferences for building public housing projects in urban rather than rural areas. For years, urban consumers received cheap foodstuffs at the expense of small-scale rural producers; migration to the city was one result.

Clearly a redefinition of the priorities of the development model did occur during the 1980s. The import-substituting industrialization model with all its inbuilt contradictions was replaced by a more open trade regime and a less interventionalist role for the state. One result was the

Plate 10.2 Crowded city bus, Havana, Cuba (Photo: Rolando Pujol, South American Pictures)

slowing of metropolitan growth. Unfortunately, the new model has done little to resolve most of the problems facing the poor either in the urban or the rural areas. Indeed, falling real incomes, cut-backs in government expenditure and rising unemployment led to a severe deterioration of living conditions after 1980. Arguably the new model may bring faster economic growth in the future which will restore living standards. But the combination of free trade, privatization and the rolling back of state controls may accentuate an already horrifyingly unequal distribution of income. To debate the final outcome of this radical change is beyond the scope of this chapter. What is clear is that the final outcome will critically affect the quality of urban life. The magnitude of the changes engendered by the shift in the development model renders any response by the urban planner rather insignificant.

Further reading

Bromley R and **Gerry C** (1979) *Casual Work and Poverty in Third World Cities.* John Wiley, Chichester. (Includes several chapters about the employment situation in Latin American cities. Discusses the relation-ships between poor urban workers and the 'formal', large-scale sector.)

Butterworth D and **Chance I K** (1981) *Latin American Urbanization.* Cambridge University Press, Cambridge. (Has strong sections on migration to the city and the adaptation of migrants to the urban environment.)

Castells M (1983) *The City and the Grassroots.* Edward Arnold, London. (Includes four chapters on Latin America which presents the latest thinking of one of the most influential writers on urban matters. It is also one of his more clearly written contributions.)

Collier D (1976) *Squatters and Oligarchs, Authoritarian Rule and Policy Change in Peru.* Johns Hopkins University Press, Baltimore. (Account of the policy of successive Peruvian governments to the creation of barriadas in Lima. An interesting and provocative account.)

de Soto H (1989) *The Other Path.* I.B. Taurus, London. (Provocative account of how to improve life for those living and working in the informal sector; admired by some, scorned by others.)

Gilbert A G (1994) *The Latin American City.* Latin American Bureau, London. (An attempy to provide a lively overview for the layperson of recent changes in the Latin American city. The book contains chapters on employment, housing, migration, urban management and urban protest and the future of the city.)

Gilbert A G and **Ward P M** (1985) *Housing, the State and the Poor: Policy and Practice in Three Latin American Cities.* CUP, Cambridge. (Detailed case studies of the land and housing markets in Bogotá, Colombia, Mexico City, and Valencia, Venezuela. The book includes an extensive discussion of the role of the state and public participation in housing and servicing the poor.)

Rakowski C A (ed.) (1994) *Contrapunto: the Informal Sector Debate in Latin America.* State University of New York Press, Albany, New York. (A detailed account of the debate about informal employment in Latin America including contributions from some of the authors of classic papers on the issue.)

Roberts B (1995) *The Making of Citizens: Cities of Peasants Revisited.* Edward Arnold, London. (A sociological analysis of urbanization in Latin America which is especially useful for its accounts of employment and of the emergence of urban systems during the nineteenth and twentieth centuries.)

Ward P M (ed.) (1982) *Self-help Housing: a Critique.* Mansell, London. (A series of papers debating the virtues and disadvantages of self-help housing in Third World countries, with several chapters relating specifically to Latin American experience.)

CHAPTER 11

Geopolitics in South America

Arthur Morris

The importance of geopolitics

One of the most important sets of ideas for a better understanding of contemporary Latin America is contained in geopolitics. Internationally, it can help interpret power struggles between nation-states; within individual countries it can help to understand curious phenomena such as inordinately expensive investment programmes in frontier regions, the movement of capital cities, or the declaration of free ports and industrial zones in regions with no suitable infrastructure.

Geopolitics may be described as a set of concepts which provide a scientific or pseudo-scientific rationale for the territorial expansion of nation-states, or which explain the relative strength and weaknesses of states in terms of their position, shape, topography and resources. Apart from historical explanations, the ideas can of course be used to project the future, or to make a claim for expansion to have better access to resources, better position in a continent, or simply to follow a pattern of expansion because this is seen as healthy.

Several features of Latin American countries make the geopolitical ideas of particular relevance.

1. At the time of independence, frontiers were poorly defined and often unsurveyed. This gave scope for dispute over their location. Disputed areas were often unpopulated, and detailed survey was difficult. Frontiers have proved unstable in the period since independence in the early nineteenth century. Changes through expansion, notably by Brazil, Peru and Chile, but also by other countries, have created the idea that frontiers can successfully be re-drawn.
2. During the present century, most countries of the region have, at some time, been in the hands of military governments. In 1980 for example, all states of South America had military governments except for Ecuador (military left in 1979), Colombia (military to

272

1968) and Venezuela. The military have found that geopolitics can provide a rationale for the creation and maintenance of large armies that can challenge their neighbours and protect present frontiers.

3. Some non-military governments have been undemocratic and authoritarian, and challenging outside states can be resorted to as a way of legitimizing their own existence. Geopolitical reasoning is along lines which link to the idea of the state as a separate body, with its own impulses and needs, in an international power struggle. But at the same time, the geopolitical theories endow these impulses and needs with a kind of academic and intellectual basis which may make it more acceptable.

The standard version of authoritarian states and the logic for their action views them as quite separate from the populations they govern. In Latin America, a distinctive blend has emerged which must be analysed in its own right. In this chapter we will show that the geopolitical ideology has been widely disseminated and many of its ideas accepted by most of the population concerned. In Ecuador, for example, all maps of the country are required to show the large part of Amazonia, lost to Peru in 1941, as part of the national territory. All Venezuelan maps have to show most of the Republic of Guayana as a Venezuelan territorial claim, though this land was never occupied by republican settlers or military forces. In Argentina, at the entry to small towns in Patagonia, as well as on the road from Ezeiza airport to Buenos Aires city, there are signs stating the Malvinas (Falkland Islands) belong to Argentina.

In other instances, it is harder for government to present a geopolitical policy in rational terms. Thus in Brazil, wider awareness of international alarm over the destruction of the Amazonian rainforest has made many educated Brazilians hostile to the continued state support for expansion of roads, mining camps and hydroelectric installations on the big rivers. But at least the government has been able to present its expansionist programme as a matter of regional development, rather than active occupation of a frontier zone which threatens its neighbours.

Geopolitical ideas can thus form a basis for the understanding of many events in the recent history of Latin American states. These states, created in the nineteenth century, do not have very stable frontiers, and most of them have undergone changes in the last hundred years. As long as territorial changes remain possible, there will be a role for geopolitics, which claims to provide a scientific spatial and historical rationale for territorial expansion or realignments.

It might also be claimed that there is little chance for new changes of national frontiers, that although there have been changes in the past the matter is now settled, and political behaviour more comparable to that in North America or Western Europe is expected. But this position overlooks the fact that recently most of the states were in the hands of military governments, not as temporary incursions into the political field, but on a semi-permanent basis, attempting to conduct the whole economic and social life of the country, and often using a strong doctrine involving geopolitical ideas.

Even civilian governments have frequently acted aggressively towards their neighbours, with little reference to authoritarian regimes in Latin America, which have had only a facade of democratic institutions that have not been respected in main lines of policy. Even this kind of reasoning, which assumes that the populace in a given country does not support the more aggressive stances or actions of its leaders, is commonly inaccurate. Within each country, a considerable effort has been made, over a long time, to gain public support for the geopolitical arguments, as will be shown for the Argentine case in this chapter. Geopolitical ideas are widely accepted and debated in Latin America, which is a good reason why the relevant arguments and attitudes should be discussed in this chapter.

The scope and method of geopolitics

In the cases that follow, geopolitics is seen as an element in nationalism, a nationalism which is expressed in territorial dimensions, as claims for the occupation of land on the frontiers or occupied by other states. It is this kind of geopolitical stance which is most dangerous in terms of the possibility of conflict between nations, and which is most visible to the outside world.

Geopolitics also relates to the realignment of external relations between states. Geopolitics has been used, in Argentina, as an aspect of the arguments for realignment of Argentina away from dependence on North America and Europe and towards its neighbours. Another role for geopolitics is to rationalize heavy state intervention. The state, military or civil, which seeks to define a strong role for itself and demonstrate the need for its action, can find a number of rationales; in the West, intervention has been justified in terms of the welfare state. In Latin America, another rationale has been national security and development of the regions, and geopolitical arguments have been used.

A variety of methodological approaches is used in the geopolitical literature. A great deal of writing has taken up the established Anglo-Saxon themes, of continentality versus maritime position. Whether Brazil is to be seen as a maritime power or a continental one is covered by many of the Brazilian geopoliticians, as is the equivalent argument for Argentina. Mackinder's heartland thesis, developed for the Eurasian continent, is modified and applied in new ways by such writers as Golbery in Brazil. These writers have pointed out the impenetrability of interior South America almost up to the present, the empty heart of the continent because of the poor access by rivers, and the threat posed if the heartland becomes accessible.

Another large set of ideas extends Carl Ritter's analogy of the national state with a living organism, which needs to and may be expected to grow, and then slow and decline. This analogy, which lies outside the Anglo-Saxon concept of the state as the creation of its people, reifies or gives a separate identity to the state, something with a life of its own. Having a separate identity, we can understand the plans put forward by geo-strategists such as Golbery for the territorial expansion of the effectively controlled state into Amazonia, or the triumphalism of Argentine expansion in the nineteenth century, into Patagonia.

The arguments are best exemplified from two large states, Argentina and Brazil, where huge territories have been loosely held, and substantial challenges to the status quo have been mounted.

The Argentine case

The base

An understanding of Argentine geopolitics can help us to interpret such matters as the urge to develop remote areas of Patagonia and Tierra del Fuego, to create a new capital city in Patagonia, and the Falklands War of 1982. Any analysis of the long-drawn out disputes with Chile, must also take into account geopolitical stances. International co-operation on the management of the Paraguay–Paraná river system is also conditioned by geopolitical considerations. But for a true understanding of these matters we need to turn first to the historical record.

Independence endowed the Argentines with a huge territory, but one with uncertain frontiers. The area had not been surveyed, notably along the Andes, and it was occupied in any case by Amerindian peoples. Alongside uncertain territorial definition, Argentina had a population which was tiny at the time of Independence in 1825, and which grew as a result of massive immigration from Europe in the later decades of the nineteenth century. There was a conflict in the national geopolitical identity, since the interior provinces sought a different kind of orientation from the Pampas, and this was a cause of the civil war which took place over most of the 1840s and 1850s before unification. For the Pampas and Buenos Aires, the interest was in modernization, in the development of production of grains and meat for export to Europe.

It is possible also to analyse the problem of weak national identity, from the colonial endowment. While there was a strong localism in each region of Argentina, a *caudillismo*, the following of the local *caudillo* or strongman, and also, amongst the élite, a sense of belonging to a wider, international Spanish-based culture, there was no intermediate, national level of identity. This had to be created, and through a directed educational policy, starting with the geography textbooks, the nationalism to emerge would eventually be territorial (Escude 1987; Shumway 1991).

Positive and negative territorial ideas were projected. On the positive side, there was the internal expansion of the country to conquer its own territory, from the Amerindians. This was a 'Conquest of the Desert', in 1878–82, under General Roca. A little later, there was a similar campaign which ended with the conquest of territory in the northern Chaco lands,

275

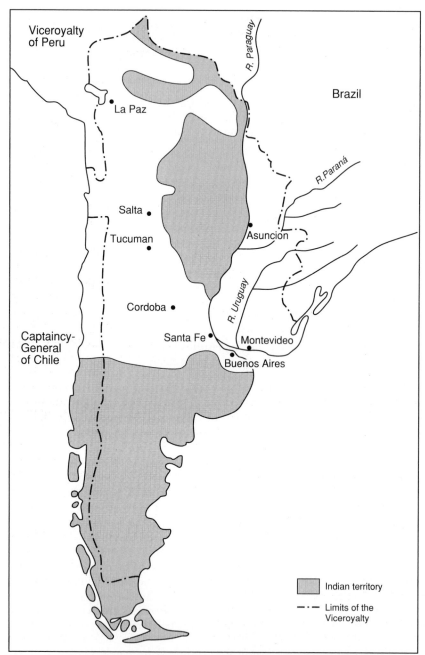

Fig. 11.1 Occupied and Indian territory in the Viceroyalty of the River Plate. The map indicates two features of importance: the territorial near-separation of the Interior from the Littoral zone of the Plate estuary and the extent of the Viceroyalty itself, all of which is claimed by Argentina as the 'inheritor' of the Viceroyalty lands.

establishing Argentine presence. This was an important movement, given the previous situation with only half the country under effective control, and a near division into two sectors, interior and pampas (Fig. 11.1).

At the same time, there was another negative territorial idea, that of the loss of Argentine territory since colonial time, so that land must be claimed back from others as rightly belonging to Argentina (Fig. 11.1). This 'loss' is from the land which formed the Viceroyalty of the River Plate in 1776, although this unit was only a device for local administration, formed late in the colonial period. This unit included Bolivia (before that country suffered heavy territorial losses), Paraguay and Uruguay, in addition to modern Argentina.

Three models of Argentina

This combination of actual territorial expansion with a sense of unsettled boundaries, and the needs for unification of a diverse immigrant population, gave good grounds for geopolitical thought exercises. Three of the spatially organized models are described here.

The earliest version of a strategic geopolitical map (Fig. 11.2) comes from Admiral Segundo R. Storni, who in 1916 presented his concept of Insular Argentina. Basing himself on the ideas of US Admiral Mahan and British geographer Halford Mackinder, he contrasted continental and maritime interests. Argentina lies in the maritime hemisphere (the half of the globe dominated by sea areas) and thus her interests are marine-oriented. Argentina is concerned about Uruguay, because of the River Plate which is shared between them. She has strategic interests in Chilean developments because they represent potential access lines to the Pacific across a short neck of land. Compared with these interests, those in Brazil are marginal, and even more so those in Bolivia and Paraguay, totally landlocked states. Storni calls attention to the South Atlantic, Argentina's own coastline, a broad front which must be used by the country, otherwise it will be threatened by others.

A second version of Argentina's position on the global stage, comes from General Juan Enrique Guglialmelli, an army commander retired in 1968, but remaining as head of a military college for officers, comparable to institutions in Brazil and elsewhere, the Escuela Superior de Guerra in Buenos Aires. Guglialmelli was also influential through being director of other institutes and editor of *Estrategia*, the leading publication on geopolitical themes of immediate interest.

Guglialmelli's model (Fig. 11.3) was of Peninsular Argentina, a country with two elements, a continental one to the north, and a peninsular one to the south. These are divided by a line approximating the Rio Negro, and a transition belt in SW Buenos Aires province, La Pampa and Neuquen (Fig. 11.3). The northern, continental belt has interests in neighbour countries to the north although physical links had not been developed. It debouches principally through the River Plate, but also has many overland links. Some of these are through Argentine Mesopotamia, the land between the Parana and Uruguay rivers,

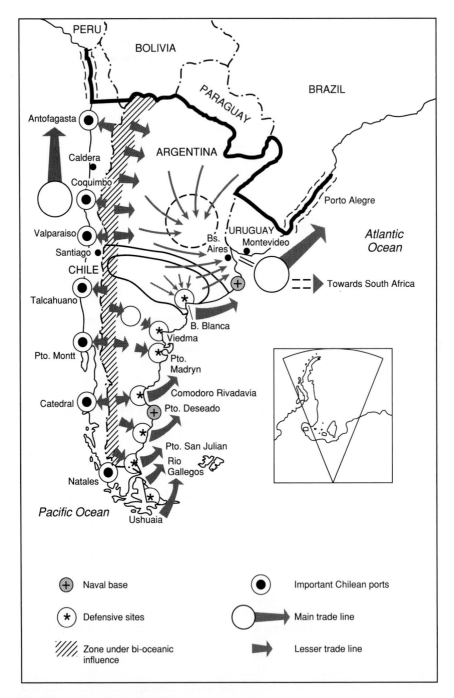

Fig. 11.2 Storni's model of Insular Argentina

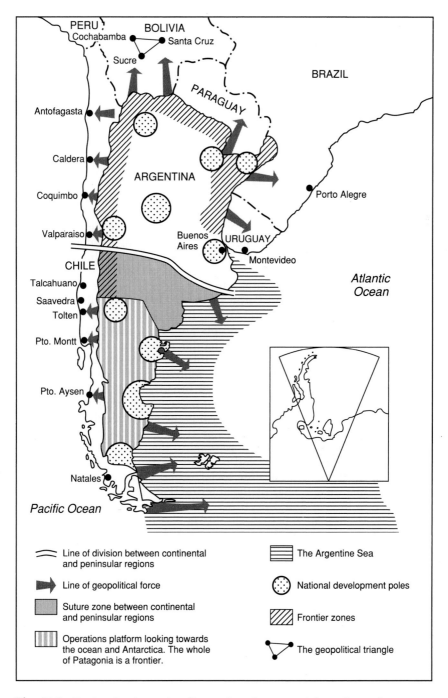

Fig. 11.3 Peninsular Argentina (*Source*: based on a map from General Guglialmelli)

towards Brazilian ports, and Guglialmelli sees this as a zone of tension between the competing influences of Brazil and Argentina.

The peninsular area looks outwards to the ocean, and the whole thin zone of Patagonia may be regarded as a frontier. For Guglialmelli, it may also be seen as an operations platform, a term which betrays the military conception of this view. This peninsular zone extends out to the ocean, including thus the Falklands, South Georgia and South Sandwich, and the Antarctic zone claimed by Argentina.

Much of the action advocated by Guglialmelli is economic, especially in the development of natural resources such as iron ore and oil and hydroelectricity. Growth poles would also be created (the Third National Plan had included them), and would be located so as to ensure development of peripheral regions. In this model, emphasis is laid on the economic independence conferred by development of resources by and for the country, liberating it from dependence, allowing the country to act on its own to defend its interests. It is a moderate statement, with little threat to any outsiders, but of course it could be interpreted in various ways.

Geopolitics was not an intellectual monopoly of the military in Argentina. Under the second Perón regime in 1973–76, civilian versions with a new idea were published. Among these were Osorio Villegas and Gustavo F. J. Cirigliano, whose concepts are here used to exemplify the era. Cirigliano's structure was of Triangular Argentina, with three main axes of importance. He contrasts the Interior axis, that of the Andes, with the existing main Littoral axis, following the River Plate, the Parana river, and linking Buenos Aires broadly with São Paulo (Fig. 11.4). A third axis is that of Patagonia, at present only lying in wait.

A principal emphasis in this portrayal is the economic dependence of Argentina on the exterior, through her trading relations with Europe and North America. This is again ascribed to the National Project of the 80s Generation and a rallying call is again made for a new project for Argentina. Cirigliano contrasts the social, economic and political content of the two main axes. That of the Interior (Fig. 11.4) starting in Talca, Chile, and ending in Quito or Tumbez, Peru, was the older axis, well developed in colonial times, especially in the tract between Cordoba, Salta and Lima. This was the densely peopled axis, and the line which would have been developed by San Martin and Bolivar, the liberators of South America who sought a United States of South America. This project had failed through the balkanization of the continent, aided and abetted by European powers whose interests were best served by small and weak South American states. Development of the Interior axis should be of a socialist character, emphasizing community relations and co-operation. This fits in with the socialist self-portrayal of the Perón government at this time.

In contrast, the Fluvial or Littoral axis is one developed in modern times, in relation to the capitalist world, externally oriented, and relying on private initiative and capitalism. Its development is autonomous, and does not require government intervention or stimulation. This development should not be stopped, but redirection of state intervention

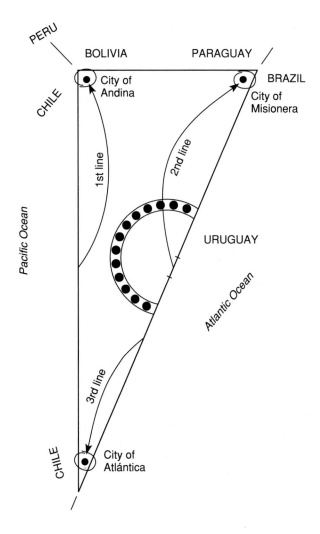

Fig. 11.4   Triangular Argentina – the clearest expression of a Peronist view of geopolitics. At each apex of the triangle, a city of over 1 million inhabitants, acting as links to the exterior, will be created. The cities are Andina, Misionera and Atlántica

should be away from the great metropolises and towards the Interior. Development of the Interior axis has actually been checked by the Littoral axis, and resources and population flowed out of the Interior to the Littoral.

Cirigliano's third axis is that of the South Atlantic, the Patagonian front. This was not entirely a new axis – it had been embarked on by Rosas in the 1830s, and by Roca in the Conquest of the Desert campaign of 1879–80. At the present, it needs further development with special attention to the ports and to the raw materials it may provide.

Another aspect of Triangular Argentina was the need to emphasize the vertices of the triangle. This meant the development under state tutelage, of three great cities, one on the Bolivian frontier, named Andina, another in the north-east corner, called Misionera, and another in Tierra del Fuego, given the name Atlántica. A play on words is made by Cirigliano. If the 80s Generation project was the port, the *puerto*, then the modern one should be the doors to Latin America, the *puertas*. Each of these great cities, the Argiropolis, would be created by the government, and each is conceived as 1 million inhabitants in size, to have sufficient economic and social size to ensure their survival and prosperity. These cities would be populated by native Argentines, and by many immigrants from Latin America, bolstering the Latin identity of the cities and of the country. Villegas identified even more clearly than Cirigliano the role of the immigrants; they would serve to bolster Argentine population, within the context of a major population policy that had to face up to the prospect of a Brazil whose population at the end of the century would reach 200 millions. Argentina needed to populate the country, and especially the remote regions.

This immigration aspect may not have been widely accepted. If we look at the writings of yet another geopolitical writer, Guillermo Terrera, we have the outline of what was probably a more typical Argentine view of ethnicity (Terrera 1983). This expresses the fear of excessive immigration, and the pressure implied by the over-populated neighbour countries. In southern Chile, it contrasts 8 million Chileans with 800,000 Argentines in Patagonia, and fears that there may already be half a million Chileans in Patagonia, mostly without documentation. Bolivians, Paraguayans and, in the future, Brazilians, may be seen as potential invasion of humanity which Argentina should check. But Terrera's comments do not detract from the force of Cirigliano's argument for a grand strategy.

Cirigliano has views about another grand fantasy, the removal of the capital from Buenos Aires, outward-facing and representing hated colonial links, to the interior. In place of that project, later to be advanced again by the civilian Alfonsin government after the Falklands War, for a new capital away from Buenos Aires at Viedma in northern Patagonia, the project would establish three new capitals, each with their own infrastructure, with great universities, with links to the other regions. Finally, a compromise between the two main axes, Interior and Littoral, is suggested, converging on the meridian of 62°W. Based on this line, the interests of both external and internal lines might be reconciled. The precise content of this line is not elaborated, but its geographical course, running close to the Falkland Islands, and close also to Viedma, is perhaps an augury of the later history.

Cirigliano's contrasting of the Interior and Littoral views of national interest date back to the nineteenth century conflicts between Buenos Aires and the Interior provinces, only resolved in the 1850s after long civil wars. This conflict and its resolution in favour of the Littoral was used by various writers, among them Arturo Jauretche. He contrasted the vision of Buenos Aires city, the *patria chica*, concerned only with production for export from the Pampas and unconcerned at Argentina's loss of interior provinces to neighbour states, with the broader vision of men from the interior, who tried to maintain the *patria grande*. These statesmen saw their cause lost, through the hidden British interests, which were served well by the balkanization of the River Plate area. Traitors to the cause of *patria grande* were such men as Sarmiento, who although coming from the interior, took the Buenos Aires view on becoming president, claiming that *'el mal que aqueja la Argentina es la extension'*, i.e. Argentina's problem is its great size. One element of Peronism was thus the reawakening of the old concerns of Argentina for its interior, though Cirigliano makes no direct reference to Argentine claims on lost provinces. Box 11.1 discusses the future for Argentina.

Box 11.1   Which way does Argentina look today?

At the present, Cirigliano's ideas for linking with the neighbour Spanish-speaking republics look dubious. Old regional associations such as LAFTA, the Latin American Free Trade Association, never managed to break into a real process of integration. What has been more successful is linkage with Brazil, under the auspices of MERCOSUR. This agreement between Argentina, Brazil, Paraguay and Uruguay, signed in 1991, calls for free movement of capital, goods and services between the member states, and a common external tariff by 1995.

The advantages of linking between the two giants are obvious: Argentina produces grains and oil products which Brazil has to import; Brazil has modern industries, tropical fruits, timber and paper, steel, at lower prices than they can be produced in Argentina. The Brazilian market of 150 million is attractive to Argentine industries, much more so than the small markets of countries such as Peru or Bolivia.

There are problems; Argentines complain of Brazilian dumping of cheap paper, steel and textiles; Brazilians in Rio Grande do Sul fear competition in farm goods from the more fertile and productive Argentine Pampas. The transport network linking the countries is deficient.

But it is still likely that geopolitical stances of antagonism will be overcome by economic convenience, and MERCOSUR may achieve this. These two countries are becoming 'wrapped in a mutual embrace' (*Financial Times*, 17 January, 1994, p. 12), which could resolve the conflict that the politicians and the generals have been unable to resolve over the years.

A rationale for geopolitics

Lines of thought were developed in Argentina to back up her nationalism in a territorial manner, notably the expansion of the occupied area north and south in a glorious campaign to Patagonia and the Chaco; and the negative recording of territorial losses.

We will now examine two of the propositions used in recent times within Argentina, which add up to an ordered discourse providing a base for an aggressive territorial stance. This discourse centres on a few concepts, among which that of the 'national project', and of lost territory, are prominent.

The National Project

A first national project had been established by the Liberal government of the 1880s. At this time development became focused on opening the country to the exterior, and encouraging the export of raw materials to Europe, principally to Britain, and import of manufactures from her. This project, according to the discourse, had become inoperable from the 1930s with the breakdown of world trade, and the country was in dire need of a new one. For civilian intellectuals, the new National Project could be the Latinization of Argentina, turning her back more on Europe and North America, linking to fellow states, especially those that had moved towards radical anti-capitalist solutions, as in Bolivia after the revolution of 1952, in Cuba, and in Peru after 1968. For military thinkers, still strongly nationalistic but perhaps more aware of global strategy, the expansion of Argentine influence on the open Atlantic front was seen as a more promising project, and could be advertised to the public as a combination of national economic development and national security.

Since these two were intimately linked together, and since the military had to ensure national security, military intervention to control the state was justified, and took place in this country, not as previously had happened in Latin America, as a temporary incursion until civil disturbances had been quelled, but as a permanent governmental force. From 1966–73, under the Revolucion Argentina of President Ongania, and again from 1976–83, in the Proceso de Reorganizacion Nacional, of General Videla, then General Galtieri, the military combined internal repression of any left-wing tendencies, with a strong external aggressive stance and heavy investment in military armaments. Argentine military expenditure in the early 1980s was, at 17 per cent of state expenditure, the highest in Latin America.

There are obvious flaws in the arguments above: it is not possible to demonstrate the need for a military intervention, when other, more successful countries had been able to progress without such an action. Military intervention in any case negated democratic will and reduced Argentina's claim to be an upholder of western cultural and political traditions. But the need for a formal national project is itself suspect; what seems to be proposed is instead a territorial form of nationalism, creating national feeling around an expansionist territorial claim.

Loss of territory and irrendentism

The other proposition mentioned above as a base for strong geopolitics, was the loss of territory, or more currently, the image of loss of territory; the counterpart for such loss is a claim for its return, called irredentism. This begins, for all writers in this vein, with the dismemberment of the Viceroyalty of the River Plate, at the time of the wars of Independence. Because this Viceroyalty had included, from its formation in 1776, Paraguay, Bolivia and Uruguay, along with Argentina, the new Republic of Argentina was considered to have lost these regions.

But there are flaws with this argument as much as with the previous one, and here on very factual grounds. There was never any formal recognition by Spain or any other country that Argentina was the successor state to the Viceroyalty of the River Plate. Some parts of Argentina were not even part of the Viceroyalty but taken from Indian occupation, specifically the area of Patagonia. If there was any logical claim to Patagonia, it was in fact from Chile, since the Captaincy General of Chile was stated to range 100 leagues (346 miles) inland from the Pacific coast. Cuyo and Tucuman, further north, had been formally taken over by the Viceroyalty of the River Plate, but Patagonia was Indian country. The strong Argentine belief in a continuous loss of territory is due to a deficient, and deliberate, educational policy, centred on the geography texts used in Argentina over the past 50 years (see Box 11.2).

Box 11.2 Questionnaire replies on territorial losses by Argentina

A questionnaire commissioned by Carlos Escude, and attached to a Gallup poll in Argentina in March 1985, asked whether the respondents believed Argentina to have lost territory throughout her history.

Answers were 73 per cent affirmative. In the breakdown of this answer, of those with tertiary education 86.1 per cent believed this to be true, while only 61 per cent of those with incomplete primary education.

In fact, in 1811, Belgrano recognized part of Misiones province as belonging to Paraguay, but this was later taken back. In 1852, the whole province of Formosa had been recognized as Paraguayan. In the colonial period, no map of Spain showed Patagonia as belonging to her.

Source: based on Carlos Escude (1987): 102–3.

The South Atlantic conflict

In April 1982 Argentina and the United Kingdom went to war over the Falklands/Malvinas Islands. This event is variously interpreted. There was first a misunderstanding at diplomatic level about intentions on either side and, more fundamentally, about the determination of the UK to defend the islands. It is also obvious that the Falklands attack was a device used by the President, General Galtieri, at a time of considerable internal problems for the government, because of the Dirty War to eliminate supposed Communist terrorism, and because the economy was in dire straits. Taking over the islands would redirect national attention on an outside project around which all interests would converge. Yet another factor was the resource .implication, the fact that this area of the South Atlantic held large marine resources in the krill and fish of the continental shelf, and potentially large resources also of oil under the shallow seas between the Falklands and the mainland. Occupation of the Malvinas was of course not to be seen in isolation, since the British themselves had designated South Georgia and the South Sandwich Islands, together with the segment of Antarctica claimed by Britain, as the Falklands Islands Dependencies, putting the whole as one package. Thus there was a large territorial gain to be anticipated.

Following our previous discussion of Argentine geopolitics, none of these explanations seems adequate, although there are elements of truth in each. There was misunderstanding; there was a need for Galtieri to 'do something' to inspire new confidence in his government; there are recognized resources in the marine area round the Falklands. But a long-term commitment to action on this Patagonian front was always in the minds of Argentine governments, and accepted by the population, which had been educated to the idea that the Malvinas had always been Argentine territory, and that this was one area in which the nation had lost territory to the outsiders. It is important to note this popular acceptance of the government's actions, despite it being a military geopolitical proposition outlined above – a national project, and ir-rendentism/loss of territory, were positively involved in the fight for the Malvinas, and so the people especially in Buenos Aires, were generally supportive of the action. It is important to note the support of Buenos Aires; these were the more educated people, with more information, and the conclusion must be that these people had been subject to more propaganda in favour of action. It was of course important to be able to show on television and other media the strong support in the nation's capital city. All the geopoliticians had accepted this, and action to recover the islands would be the satisfactory conclusion of an historical episode. War to regain the Malvinas can thus be seen as simply part of a well-understood national strategy which involved movement on a Third Front, out to the South Atlantic. It is also part of a national search for identity, which is how the 'national project' arguments may be interpreted, a constant battle to bring together demographic elements and regional interests which have never truly coagulated.

The position of Argentina's claim to rightful ownership of the islands, now and at the time of invasion of the islands, was accepted by virtually all the population. Parts of a national project had entered into national education, which has been standardized at least since the time of Perón's first period in power, to include a version of the claim. The grounds for ownership are so different from those of Britain that agreement or a common understanding are almost impossible without a damaging change in the story. The Argentine version is given, with only minor variations, in all secondary school geography texts (see for example, Daus 1950; Sarrailh *et al.* 1988), whose content list has been controlled by central government to the present.

At an academic level, some writers have sought to understand the context in which a Malvinas war could be launched; the nature of the state that could conceive such a project. Many of the analyses have concentrated on the outside influence, that of the USA, inculcating ideas of national security in the post-Vietnam era, and the linking of economic development with national security in a giant project. These ideas have been followed also in the presentation above. However, this is an oversimplification of reality. There are important social divisions in Latin American military forces, and the officers represent an élite group of society, whose expression may have been repressed under some recent democratizing regimes and who have used military power to find this expression. An authoritarian state is a natural situation given the structure of society in some developing countries, and the military are merely one form of this authoritarian state.

Summarizing the causal links seen in the Argentine case, we may say that there are three elements in the chain, a social and political structure, an intellectual construct, and an immediate stimulus. The basic structure is the authoritarian state, reinforced by selective adoption of ideas from abroad. The intellectual construct is the geopolitics, and the immediate stimulus is the opportunity seen to take the islands.

Brazil

In Brazil in recent years, under the influence of world opinion, there has been much heated debate over the issues of rainforest depletion, and its effect on the global ecological balance, the greenhouse effect, climatic change and the loss to science of unknown species. Another argument has been over the conservation, not of nature, but of primitive peoples, the Amerindian tribal groups. Large numbers of these have been killed by agents of ranchers anxious to clear their property. Others have contracted Western diseases; still others lost their livelihood through the flooding of territories by the building of dams, or through destruction of habitat for game in road building and mining operations.

Box 11.3 The Calha Norte

Calha Norte Indigenista Missionario, in Manaus, set out a paper against the project to develop the whole northern frontier zone, Calha Norte, in 1987:

'The Calha Norte project is of a belico-military and developmentalist nature. It affects a strip of frontier 6,500 km long, north of the Amazon river, extending to 14 per cent of national territory, and affecting five neighbour countries.'

'The philosophy behind Calha Norte is preoccupying. One has to read the justificatory statements given for the Programme, to understand the real consequences for local destruction of regional resources, and the marginalization of the inhabitants, which we have already seen through other authoritarian and anti-people projects, such as Jari, Carajas, Tucurui, and Polonoreste. The construction of the Transamazonica road signified the ethnocide of thousands of Indians. ...

In fact Amazonia has always been and continues to be a colony within Brazil.'

'The Calha Norte project will cause the disappearance of native peoples in the north of the country. And who will preserve us from the other "Calhas", to south, to west, or to east?'

For ordinary Brazilians, the occupation of the interior, or of Amazonia in particular, the most strongly contested operation, is not just a matter of geopolitics, it is a vital part of being Brazilian (see Box 11.3). In this country, perhaps the most important part of an expansionist strategy is the occupation of frontier areas by farmers, in a centuries old pattern of occupation of new lands, their exploitation for a few years, followed by abandonment and pushing on to the west. Some of them have occupied areas even beyond the present boundaries of Brazil, in eastern Paraguay and eastern Bolivia.

At another level, educated Brazilians are able to rationalize expansion of infrastructure and deforestation in the interior in terms of the need for regional development. Since the southern and south-eastern states have much higher living levels than the rest, this seems a rational presentation. Examined a little more closely, it stands in contradiction to the observed process of mass migration of poverty-stricken people from the north-east to São Paulo and Rio, a process still going on and little helped by regional development programmes to date.

On the other hand, the failed attempts to colonize Amazonia associated with the Trans-Amazonica Highway, the massive investments in hydroelectric schemes such as Itaipu, the incentives given to foreign firms to invest in unproductive ranches in the north instead of paying this money as federal taxes are all intelligible when we consider the hidden agenda, in which the actual occupation and domination of territory and resources is itself seen as a good.

There are many points of comparison between the cases of Argentina and Brazil. In both cases, there has been a large internal 'empire', the unsettled interior of the country, to be conquered. In the Brazilian case, this empire was more to be conquered from nature than from any hostile group. In each case, too, there have been tendencies towards fission and the need for a centralizing project around which to organize the state. Brazil, like Argentina, has been outward-looking, its cultural, political and economic norms related to Europe more than to its own interior in the last century. As in Argentina, Brazilian writers have been in the habit of using space creatively, concerned with the location of a national capital in the interior, with changing regional emphases.

One big difference has been the greater and more formal use of the several intellectual models of geopolitics, including those coming direct from the German school of Haushofer, emanating ultimately from nineteenth century writers. Whereas Argentine geostrategies have had to show some invention in their emphasis of the different fronts, Brazilians could feed directly on the concepts of Admiral Mahan and Halford Mackinder, regarding a continental versus a maritime position, and the fate of the states enjoying either of these. This Manichaean vision allowed some of them to emphasize the importance of a central position for Brazil within the continent. More importantly, Ratzel's comparison of the state to an organism was picked up and developed by the Brazilians as a useful concept to justify aggressive expansion.

Early statements on geopolitics

One of the first Brazilian writers to enunciate geopolitical statements was Everardo Backheuser. He sees the state as comparable to a tree, with its roots its geographical territory, which must be an essential part of the state organism. Backheuser emphasizes the need to promote national feeling and combat regionalism, but rejects the idea that the military are the only group capable of generating this feeling. His ideas, however, seem to have been widely studied among the military in later years.

In the 1930s, Mario Travassos was the best-known geopolitical writer. He identified the central zone of South America as containing two antagonisms, between Atlantic and Pacific, and between River Plate and Amazonia. These intersect in Bolivia, in the zone of Santa Cruz, which is therefore a dangerous power vacuum. Travassos discounted the importance of the centrality of the Mato Grosso and Goias plateau of Brazil.

In the post-1945 era, a set of new factors made geopolitical ideas more relevant to Brazilian politics, and the whole geopolitical paradigm is usually portrayed as dating from this period of Brazilian history, though there are pre-1939 precedents. A central figure was Golbery do Couto e Silva (generally referred to as General Golbery). He constructed a new version of Brazilian geopolitics. It was based partly on the old pre-1939 ideas of centrality, continental dominance and control, but developed into a concept akin to a military operation, with theatres of action, areas to be advanced into, and a broad scope to the actions taken.

Golbery had an extraordinary influence on the growth in stature of geopolitics in Brazil, due to his own energy and active support of his ideas, and to his position, as the head of the Escola Superior da Guerra, the Military College which was used to disseminate his ideas very widely among the military élite, but also among businessmen and politicians in Brazil generally. Some of his ideas were enunciated during the 1950s and others in the 1960s and 1970s, a period of military government when he was able to have direct effect on national policy. The spatial model he proposed in 1952 (Fig. 11.5) was based on the expansion of the effective national territory of Brazil, through its incorporation, in stages, into the developed area with roads and communications.

The first stage is the unification of the central nucleus with the north-east and south regions. Land links were needed. A second stage was the movement into the interior, into the centre west region. A final stage was the incorporation of the Amazon basin to become an effective part of national territory.

In this spatial model, Brazil's aims are seen as the innocent occupation of its national territory. But there are continuing references to the neighbouring countries, and another spatial pattern is that covering the whole continent of South America (Fig. 11.6). In this 1959 model, the urban-industrial core of Rio–Belo Horizonte–São Paulo is still seen as

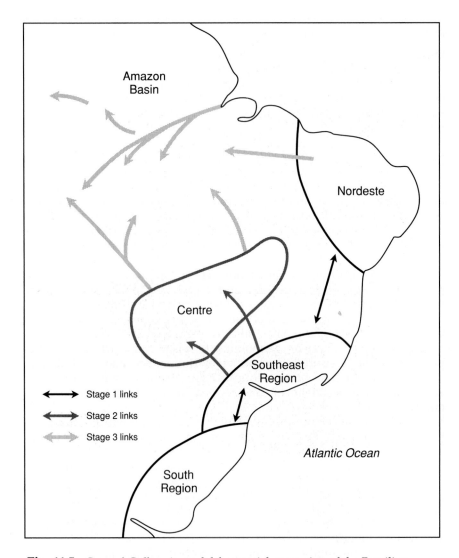

Fig. 11.5 General Golbery's model for spatial expansion of the Brazilian oecumene (*Source:* based on do Couto e Silva 1967, p. 46)

the central manoeuvring platform, but now no longer of Brazil, but of the continent. Two areas are significantly left blank on Golbery's original sketch, the area of Bolivia–Paraguay–Mato Grosso–Rondônia, and that of the Brazilian north-east. There are two macro-regions seen as weak areas, without strong definition. In particular, the Bolivia–Paraguay– Mato Grosso–Rondônia macro-region is described as a geopolitical 'welding zone', of ambivalent character poised between Atlantic and Pacific influences, and between the Plate and Amazon drainage basins. Seen in these terms, it is evident that control of this

290

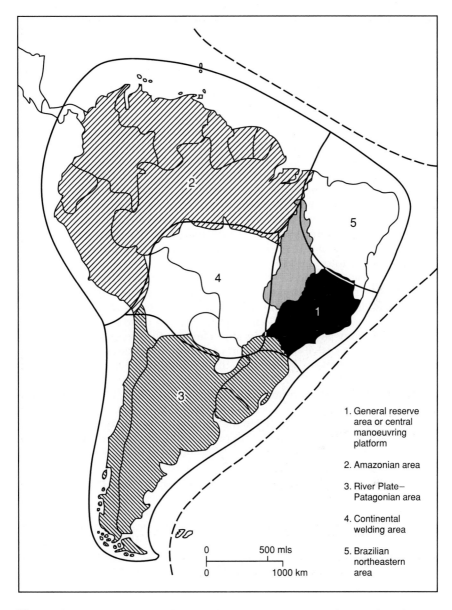

Fig. 11.6 A strategic view of the South American continent (*Source*: based on do Couto e Silva 1967, p. 90)

central belt is of vital importance; and this means control of Bolivia and Paraguay, notoriously poor, weak states, obvious buffer states between Argentina and Brazil.

Brazilian expansionism in practice

What are the factual contexts of the fertile paper constructions of Golbery, and indeed of the whole Brazilian school? How is it possible to understand recent events on the basis of this thought? How do the actions differ from those of Argentina?

The most obvious differences are that Argentina has been involved in a war, the Falklands/Malvinas campaign, as a result of military adherence to the doctrine. Brazil, by contrast, has been able to act without military force against an external enemy, as indeed has been the case in most of Brazilian history. The irony of this situation is that while Brazil has had written for it a powerful quasi-military geopolitical project, real expansion of the state, or rather of Brazilian occupation and influence over territory, has been without military action; ethnocide of Indians and widespread killing of non-Indian settlers in the way of big land-owners in some areas has been effectively accomplished without formal military intervention.

Most of the cases referred to in this chapter are direct geopolitical conflicts, even if economic interests appear relevant. More complex situations exist in areas where a demographic or economic interest exists and gives rise to a territorial claim or limitation of sovereignty. Parts of eastern Paraguay are of considerable concern to Paraguayan authority, because of the massive immigration of Brazilians in association with the expansion of soyabean cultivation. In some areas of eastern Paraguay, Portuguese-speaking Brazilian people are more numerous than the Paraguayans. Perhaps more significant still in the long run is the challenge to Paraguayan sovereignty from the joint operation with Brazil of the Itaipu power from the Paraná river. Paraguay has never had enough funds to invest in this project, nor the need to use its share of the power output. Brazil has been the obvious winner of power and influence, and has made Paraguay dependent on it through stipulating Paraguayan payment for its 'share' of Itaipu, by the transfer of the power to it at low prices.

It is logical to conclude that there are two kinds of geopolitics, one that is real and involves the expansion of influence and power in South America, which is now happening for Brazil; and the other, which has little chance of affecting the power and serves a different purpose, a kind of territorial nationalism, which we have described in the case of Argentina.

Further reading

Backheuser E (1926) *A Estructura Politica do Brasil.* Machado Editores, Rio de Janeiro.
Backheuser E (1933) *Problemas do Brasil: estructura geopolitica.* Grupo Editor Umnia, Rio de Janeiro.
Child J (1985) *Geopolitics and Conflict in South America.* Praeger, New York.

Cirigliano G F J (1975) *Argentina Triangular: geopolitica y proyecto nacional*. Humanitas, Buenos Aires. (Written from the point of view of Peronism, and based on the theme of the need to delink from the major industrial powers, and to reorientate Argentina towards her Latin neighbours. Geopolitical arguments are used in a new way here.)

do Couto e Silva G (1967) in **Olympio J** (ed.) *Geopolitica do Brasil*. Rio de Janeiro. (Formed out of a series of essays written at different times, but covering the same field. General Golbery demonstrates with brilliant scholarship the changing role of space in world politics and the need to adapt. He makes a strong case for the intimate relating of politics and geography and for the need to construct policy with geopolitical questions in mind.)

Dodds K-J (1993) 'Geopolitics, cartography and the state in South America', *Political Geography*, **12/4**, 361–81. (This wide-ranging paper links the ideas of territorial nationalism with geopolitical theories and propositions in South America, especially Argentina and Brazil.)

Escude C (1987) *Patologia del Nacionalismo; el caso argentino*. Editorial Tessis/Instituto Torcuato di Tella. (A series of essays on related themes by Escude. Written by an exceptional Argentine scholar who is highly critical of the Argentine presentation of geopolitical arguments, he shows how these arguments have been adopted into the national culture through their presentation in school texts over the past century.)

Hepple L W (1991) 'Metaphor, geopolitical discourse and the military in South America', in **Barnes T J** and **Duncan J S** (eds.) *Writing Worlds: Discourse, Text and Metaphor in the Representation of Landscape*, Ch. 9, pp. 136–54. Routledge, London/New York. (Discusses the use of the organism analogy by South American geopoliticians.)

Kelly P and **Child J** (eds.) (1988) *Geopolitics of the Southern Cone and Antarctica*, Lynne Rienner, Boulder/London. (An edited volume covering a wide variety of issues and explaining them for non-specialist readerships.)

Mattos C M (1980) *Uma geopolitica Pan-Amazonica, Biblioteca do Exercito*. Biblioteca do Exercito, Rio de Janeiro. (Updating Golbery's thought into the 1970s when the role of Amazonia and issues such as conservation became important.)

Moneta C J, Lopez E and **Romero A** (1985), *La Reforma Militar*. Legasa, Buenos Aires. (Examines the problems faced by Argentina post-Malvinas/Falklands, in setting new aims and roles for the military, and the nature of the relation of state to military in the period leading up to that war.)

Pacheco de Oliveira Filho J (1990) 'Frontier security and the new indigenism: nature and origins of the Calha Norte Project', in **Goodman D** and **Hall A** (eds.) *The Future of Amazonia*, Ch. 7, pp.155–76. St Martin's Press, New York.

Shumway N (1991) *The Invention of Argentina*. University of California, Berkeley. (A historian's view, showing the need within the country to identify national causes beyond Independence.)

Travassos M (1936) *Projecao Continental do Brasil*, Editora Nacional, Edição Brasiliana, Rio de Janeiro (originally published in 1931 as *Aspectos Geograficos Sul-Americanos*).

References

Ainstein L (1995) 'Buenos Aires: a case of deepening polarisation', in **Gilbert A G** (ed.) *Latin America's Megacities*. United Nations University Press, Tokyo, forthcoming.

Amato P (1968) *An Analysis of the Changing Patterns of Elite Residential Areas in Bogotá, Colombia*, Dissertation Series No. 7. Cornell University, Latin American Studies Program, Ithaca, NY.

Auty R M (1993) *Sustaining Development in Mineral Economies: The Resource Curse Thesis*. Routledge, London.

Barber R T and **Chavez F P** (1983) 'Biological consequences of El Niño', *Science* **22** (4269): 1203–10.

Birkbeck C (1978) 'Self-employed proletarians in an informal factory: the case of Cali's garbage dump', *World Development* **6**, 1173–85.

Borah W and **Cook S F** (1963) 'The aboriginal population of central Mexico on the eve of Spanish conquest', *Ibero-Americana* **45**. University of California, Berkeley and Los Angeles.

Briceño A, Pasco-Font A, Escobal J and **Rodríguez J** (1992) *Gestión pública y distribución de ingresos: tres estudios de caso para la economía peruana*. Serie Documentos do Trabajo No. 115. Banco Interamericano de Desarrollo, Washington, DC.

Bromley R (1979) *The Urban Informal Sector: Critical Perspectives on Employment and Housing Policies*. Pergamon Press, Oxford.

Carlstein T (1982) *Time, Resources, Society and Ecology*, Studies in Human Geography No. 49. Department of Geography, University of Lund, Lund.

Caviedes C N and **Fitz T J** (1992) The Peru–Chile Eastern Pacific fisheries and climatic oscillation,' in **Glantz M H** (ed.) *Climatic Variability, Climate Change and Fisheries*. CUP, Cambridge.

CEPAL (1982) *Notas sobre la economia y el desarrollo de América Latina*, No. 372, November/December.

CEPAL (1985) *Notas sobre la economía y el desarrollo de América Latina*, Nos. 424/425, December.

CEPAL (1991) *La equidad en el panorama social de América Latina durante los años ochenta*. CEPAL, Santiago.

CEPAL (1992a) *Gasto social y equidad en América Latina*. CEPAL, Santiago.

CEPAL (1992b) *Notas sobre la economía ye el desarrollo de América Latina, Nos. 537/538, December.*

CEPAL (1993) *Notas sobre la economía y el desarrollo de América Latina*, Nos. 552/553, December.

Cirigliano G F J (1975) *Argentina Triangular: geopolitica y proyecto nacional*. Humanitas, Buenos Aires.

Collier *et al.* (1992) *The Cambridge Encyclopedia of Latin America*. Cambridge University Press, Cambridge.

CONAPO (Consejo Nacional de Población) (1992) *La zona metropolitana de la Ciudad de México: problemática actual y perspectivas demográficas y urbanas*. CONAPO, Mexico City.

Coe M D (1964) 'The chinampas of Mexico', *Scientific American* **211**: 90–98.

Cook N D (1965) 'La población indígena en el Perú colonial', *Anuario del Instituto de Investigaciones Históricas* **8**: 73–110.

Cook N D (1981) *Demographic Collapse: Indian Peru, 1520–1620*. CUP, Cambridge.

Cordera Campos R and **González Tiburcio E** (1991) 'Crisis and transition in the Mexican economy', in **González M** and **Escobar A** (eds.) *Social Responses to Mexico's Economic Crisis of the 1980s*, pp. 19–56. University of California, San Diego.

Cornelius W A (1977) 'Introduction', in **Cornelius W A** and **Kemper R V** (1977) 'Metropolitan Latin America', *Latin American Urban Research 6*, pp. 7–24. (Sage Publications, London.)

Cornelius W A (1979) 'Migration to the US. The view from rural communities', *Development Digest* **17**, 90–101.

Cornelius W A and **Kemper R V** (1977) 'Metropolitan Latin America', *Latin American Urban Research 6*.

Crossley J C (1983) 'The River Plate Republics', in **Blakemore H** and **Smith C T** (eds.) *Latin America: Geographical Perspectives*, pp. 383–455. Methuen, London.

Cummings B J (1990) *Dam the Rivers, Damn the People*. Earthscan, London.

Dargie T C D and **Furley P A** (1994) 'Monitoring change in land use and the environment', in **Furley P A** (ed.) *The Forest Frontier: Settlement and Change in Brazilian Roraima*. Routledge, London, pp. 68–85.

Daus F A (1950) *Geografia de la Republica Argentina: Parte Humana*. Angel Estrada, Buenos Aires.

Denevan W M (1992a) 'The pristine myth: the landscape of the Americas in 1492', *Annals of the Association of American Geographers* **82**(3), 369–85.

Denevan W M (ed.) (1992b) *The Native Population of the Americas in 1492*. 2nd ed. University of Wisconsin, Madison.

Diaz H F and **Markgraf V** (eds.) (1992) *El Niño: Historical and Paleoclimatic Aspects of the Southern Oscillation*. CUP, Cambridge.

Dobyns H F (1966) 'Estimating Aboriginal population', *Current Anthropology* **7**, 395–449.

do Couto e Silva G (1967) in **Ulympio J** (ed.) *Geopolitica do Brasil*. Rio de Janeiro.

Donkin R A (1979) *Agricultural Terracing in the Aboriginal New World*, Viking Fund Publications in Anthropology No. 56. University of Arizona Press, Tucson.

ECLAC (United Nations Economic Commission for Latin America and the Caribbean) (1989) *Preliminary Overview of the Economy of Latin America and the Caribbean 1989*. ECLAC, Santiago, Chile.

Enfield D B (1992) 'Deepening thermocline and reduction of nutrient supply to the photic zone: historical and prehistorical overview of El Nino and the Southern Oscillation', in **Diaz H F** and **Markgraf V** (eds) *El Niño: Historical and Paleoclimatic Aspects of the Southern Oscillation*. Cambridge University Press **5**: 95–117.

Escude C (1987) *Patología del Nacionalismo; el caso argentino*. Editorial Tessis/Instituto Torcuato di Tella.

FAO (1992) *Forest Resources Assessment 1990*. FAO, Rome.

FAO (1993) *Forest Resources Assessment 1990: Tropical Countries, Forest Paper 112*. FAO, Rome.

FAO/UNEP (1983) *Guidelines for the Control of Soil Degradation*. FAO, Rome.

Faron L C (1967) 'A history of agricultural production and local organization in the Chancay valley, Peru', in **Steward J H** (ed.) *Contemporary Change in Traditional Societies*, pp. 229–94. University of Illinois Press, Urbana.

Fearnside P M (1990) 'Environmental destruction in the Brazilian Amazon', in **Goodman D** and **Hall A** (eds) *The Future of Amazonia*. Macmillan, London, pp. 179–225.

Frank A G (1967) *Capitalism and Underdevelopment in Latin America*. Monthly Review Press, New York.

Furley P A (1990) 'The nature and sustainability of Brazilian Amazon soils, in **Goodman D** and **Hall A** (eds) *The Future of Amazonia*. Macmillan, London, pp. 309–59.

Furley P A and **Crosbie A J** (1974) *Geography of Belize*. Collins, London.

Furley P A and **Newey W W** (1983) *Geography of the Biosphere*. Butterworth, London.

Furley P A and **Ratter J A** (eds) (1992) *Mangrove Distribution, Vulnerability and Management in Central America*. ODA–OFI Forestry Research Programme R 4736, Forestry Research Institute, Oxford.

Furley P A, Proctor J, and **Ratter J** (eds) (1992) *Nature and Dynamics of Forest-Savanna Boundaries*. Chapman and Hall, London.

Gasparini G (1981) 'The present significance of the architecture of the past', in **Segre R** (ed.) *Latin America in its Architecture*. Ch. 3, pp. 77–104. Holmes and Meier, New York.

Gilbert A G (1983) 'The tenants of self-help housing: choice and constraint in the housing market', *Development and Change* **14**: 449–77.

Gilbert A G (1993) *In Search of a Home: Rental and Shared Housing in Latin America*. UCL Press, London.

Gilbert A G (1994) *The Latin American City*. Latin American Bureau, London.

Gilbert A G and **Goodman D E** (1976) 'Regional income disparities and economic development: a critique', in **Gilbert A G** (ed.) *Development Planning and Spatial Structure*. Ch. 6. pp. 113–42. John Wiley, Chichester.

Glantz M H and **Thompson J D** (eds) (1981) *Resource Management and Environmental Uncertainty: Lessons from Coastal Upwelling Fisheries*. Wiley, New York.

Glynn P W (ed.) (1990) *Global Ecological Consequences of the 1982–83 El Niño Southern Oscillation*. Elsevier, Amsterdam.

Goldsmith E and **Hildyard N** (1984, 1986) *The Social and Environmental Effects of Large Dams*, Vol. 1. *Overview*, Vol 2. *Case Studies*. Wadebridge Ecological Centre, Cornwall.

Goldsmith W W and **Wilson R** (1991) 'Poverty and distorted industrialization in the Brazilian northeast', *World Development* **19**: 435–55.

González de la Rocha M (1988) 'Economic crisis, domestic reorganization and women's work in Guadalajara, Mexico', *Bulletin of Latin American Research* **7**(2), 207–23.

González de la Rocha M (1990) 'Family well-being, food consumption, and survival strategies during Mexico's economic crisis', in **González M** and **Escobar A** (eds) *Social Responses to Mexico's Economic Crisis of the 1980s*, pp. 115–28. University of California, San Diego.

Government of Chile (1980) 'Desertification in the Region of Coquimbo, Chile', in UNESCO/UNEP/UNDP *Case Studies in Desertification*. UNESCO, Paris, pp. 52–114.

Grainger A (1990) *The Threatening Desert*. Earthscan, London.

Grosh M (1990) *Social Spending in Latin America. The Story of the 1980s*. World Bank Discussion Papers No. 106. World Bank, Washington, DC.

Gwynne R N (1985) *Industrialisation and Urbanisation in Latin America*. Croom Helm, London.

Hägerstrand T (1957) 'Migration and area', in **Hannerberg D** (ed.) *Migration in Sweden* (Lund Series in Geography, series B) **13**, 25–128.

Hakkert R and **Goza F W** (1989) 'The demographic consequences of austerity in Latin America', in **Canak W L** (ed.) *Lost Promises: Debt, Austerity, and Development in Latin America*, pp. 69–97. Westview Press, Boulder.

Hall A (1989) *Developing Amazonia*. Manchester University Press, Manchester.

Hanson S G (1938) *Argentine Meat and the British Market*. Stanford University Press, Stanford.

Hardoy J E (1982) 'The building of Latin American cities', in **Gilbert A G, Hardoy J E** and **Ramirez R** (eds) *Urbanization in Contemporary Latin America*. Ch. 2, pp. 19–34. John Wiley, London.

Henderson-Sellers A (1987) 'Effects in change of land use on climate in the humid tropics', in **Dickenson R E** (ed.) *The Geophysiology of Amazonia*. Wiley, New York, pp. 391–408.

Herring H (1965) *A History of Latin America*. Knopf, New York.

IBGE (1950) *Conselho de Estatistica, Brasil Censo Industrial 1950*, Vol. III, Tomo I. IBGE, Rio de Janeiro.

Illanes M A (1989) *Historia del Movimiento Social y de la Salud Publica en Chile: 1885–1920*. Colectivo do Atención Primaria, Santiago.

Katz J (ed.) (1987) *Technology Generation in Latin American Manufacturing Industries*. Macmillan, London.

Kenney M and **Florida R** (1994) 'Japanese maquiladoras: production organisation and global commodity chains', *World Development* **22**(1), 27–44.

Kusnetzoff F (1987) 'Urban and housing policies under Chile's military dictatorship 1973–1985', *Latin American Perspectives* **53**: 157–86.

La Belle T H (1986) *Non-formal Education in Latin America and the Caribbean*. Praeger, New York.

Lal R, Sanchez P A and **Cummings R W** (1986) *Land Clearing and Development in the Tropics*. A A Balkema, Boston.

Larranaga O (1992) *Macroeconomics, Income Distribution and Social Services: Bolivia During the 80s*. Cuadernos Serie Investigación No. 1–50. ILADES, Georgetown University, Santiago.

Lautier B (1990) 'Wage relationship, formal sector and employment policy in South America', *Journal of Development Studies* **26**, 278–98.

Lewis O (1963) *The Children of Sánchez: Autobiography of a Mexican Family*. Vintage Books, New York.

Lovell W G (1992) 'Heavy shadows and black night: disease and depopulation in colonial Spanish America', *Annals of the Association of American Geographers* **82**, 426–43.

Lovell W G and **Lutz C** (1994) 'Conquest and population: Maya demography in historical perspective', *Latin American Research Review* **29**(2): 133–40.

McAlister L N (1984) *Spain and Portugal in the New World, 1492–1700*. University of Minnesota Press, Minnesota.

Marquez G (1993) 'Fiscal policy and income distribution in Venezuela', in **Hausmann R** and **Rigobon R** (eds) *Government Spending and Income Distribution in Latin America*. IESA, Inter American Development Bank, Washington, DC.

Mercer Fraser Ltd (1991) *International Benefits Guidelines*. William M Mercer Companies, London.

Merrick T W and **Graham D H** (1979) *Population and Economic Development in Brazil*. Johns Hopkins University Press, Baltimore.

Mesa-Lago C (1992) *Health Care for the Poor in Latin America and the Caribbean*. Pan American Health Organization, Inter-American Foundation, Washington DC.

Messner D (1993) 'Shaping competitiveness in the Chilean wood-processing industry', *CEPAL Review* **49**, 117–37.

Mitchell B R (1983) *International Historical Statistics: The Americas and Australasia*. Macmillan, London.

Molion L C (1987) 'On the dynamic climatology of the Amazon Basin and associated rain producing mechanisms', in **Dickenson R E** (ed.) *The Geophysiology of Amazonia*. Wiley, New York, pp. 391–408.

Monosowski E (1991) 'Dams and suitable development in Brazilian Amazonia', *International Water Power and Dam Construction* **43**: 5.

Moser C (1993) 'Adjustment from below: time, the triple role in Ecuador', in **Radcliffe S** and **Westwood S** (eds) *Viva: Women and Popular Protest in Latin America*. Routledge, London.

Myers N (1984) *The Primary Source*. W W Norton, New York and London.

Nieuwolt S (1977) *Tropical Climatology.* Wiley, London.

Oldeman L R, Hakkeling R T A and Sombroek W G (1991) *World Map of the Status of Human Induced Soil Degradation* (2e) Annex S. International Soil Reference and Information Center, Wageningen, Netherlands.

Pauly D, Muck J Mendo *et al* (1989) *The Peruvian Upwelling System: Dynamics and Interactions.* Instituto del Mar del Peru, Callao, Peru.

Peattie L R (1975) 'Tertiarization and urban poverty in Latin America', *Latin American Urban Research* **5**: 109–23.

Peattie L R (1979) 'Housing policy in developing countries: Two Puzzles', *World Development* **7**: 1017–22.

Peattie L R (1987) 'An idea in good currency and how it grew: the informal sector', *World Development* **15**: 851–60.

Philander S G (1990) *El Niño, La Niña and the Southern Oscillation.* Academic Press, San Diego.

Prance G T and **Lovejoy T E** (1985) *Key Environments: Amazonia.* Pergamon Press, Oxford.

Roemer M (1985) *National Strategies for Health Care Organization.* Health Administration Press, Ann Arbor, Michigan.

Ruano Fourier A (1936) *Estudio economico de la produccion de las carnes del Rio de La Plata.* Pena y Cia, Montevideo.

Salati E (1985) 'The climatology and hydrology of Amazonia', in **Prance G T** and **Lovejoy T** (eds) *Key Environments: Amazonia.* Pergamon Press, Oxford, pp. 18–48.

Santana I and **Rathe M** (1993) *Sistemas de entrega de los servicios sociales en la República Dominicana: Una agenda para la reforma,* Serie Documentos de Trabajo No. 150. Banco Interamericano Desarrollo, Washington, DC.

de Sarrailh E E O, Andina M A, Somozo E J (1970) *Geografia del Argentina.* (fourth edition) El Ateneo, Buenos Aires.

Scarpaci J L, Pio-Infante R and **Gaete A** (1988) 'Planning residential segregation: the case of Santiago, Chile', *Urban Geography* **9**: 19–36.

Schmitz H (1993) 'Small firms and flexible specialization in developing countries', in **Spath B** (ed.) *Small Firms and Development in Latin America.* International Institute for Labour Studies, Geneva.

Seubert C E, Sanchez P A and Valverde C (1977) 'Effects of land clearance methods on soil properties of an ultisol and crop performance in the Amazon jungle of Peru', *Tropical Agriculture* **54**: 307–21.

Shipley R E (1977) 'On the outside looking in: a social history of the Porteno during the golden age of Argentine development, 1914–1930.' PhD dissertation, Rutgers University, New Brunswick, New Jersey.

Shumway N (1991) *The Invention of Argentina*. University of California, Berkeley.

Skoal D and Tucker C (1993) 'Tropical Deforestation and Habitat Fragmentation in the Amazon: Satellite Data from 1978–88', *Science* **260** (5116): 1095–1099.

Stoddart D R (1963) Effects of Hurricane Hattie on the British Honduras reefs and cays, October 30–1 1961. *Atoll Research Bulletin* No. 95. Pacific Science Board NAS–NRL, Washington DC.

Storper M (1991) *Industrialization, Economic Development and the Regional Question in the Third World*. Pion, London.

Sylva P (1986) *Gamonalismo y Lucha Campesina*. Abya-Yala, Quito.

Terrera G A (1983) *Geopolítica Argentina: población, fronteras, comunicaciones, antropología*. Editorial Plus Ultra, Buenos Aires.

Thiesenhusen W (ed.) (1989) *Searching for Agrarian Reform in Latin America*. Unwin Hyman, London.

Thomas D S G (1993) 'Sandstorm in a teacup? Understanding desertification', *Geographical Journal* **159**(3): 318–31.

Thomas D S G and Middleton N J (1994) *Desertification: exploding the myth*. Wiley, Chichester.

Tout K (1989) *Ageing in Developing Countries*. Oxford University Press/Helpage International, Oxford.

Trejos J D (1993) *Sistemas de entrega de los servicios sociales: Una agenda para la reforma en Costa Rica*. Serie Documentos de Trabajo No. 153. Banco Interamericano de Desarrollo, Washington, DC.

Trussel D (ed.) (1992) *The Social and Environmental Effects of Large Dams*, Vol. 3. *A Review of the Literature*. Wadebridge Ecological Centre, Cornwall.

Uhl C and **Kauffman J B** (1990) 'Deforestation, fire, susceptibility and potential tree responses to fire in the eastern Amazon', *Ecology* **71**(2): 437–49.

UNCED (1992)

UNDP (United Nations Development Programme) (1991) *Human Development Report 1991.* United Nations, New York.

UNDP (1992) *Human Development Report 1992.* United Nations, New York.

UNDP (1994) *Human Development Report 1994.* United Nations, New York.

UNEP (1990) 'Global assessment of soil degradation' in *World Map on Status of Human Induced Soil Degradation.* UNEP, Nairobi, Kenya.

UNEP (1992) *World Atlas of Desertification.* Edward Arnold, London.

UNEP (1993) *Desertification Control Bulletin 22*, 84 pp. United Nations Environment Programme, Nairobi, Kenya.

van den Berghe P and **Primov G** (1977) *Inequality in the Peruvian Andes: Class and Ethnicity in Cuzco.* University of Missouri Press, Columbia.

Villa M and **Rodríguez J** (1995) 'Demographic trends in Latin America's urban system, 1950–1990', in **Gilbert A G** (ed.) *Latin America's Megacities.* United Nations University Press, Tokyo, forthcoming.

Villaran F (1993) 'Small-scale industry efficiency groups in Peru', in **Spath B** (ed.) *Small Firms and Development in Latin America.* International Institute for Labour Studies, Geneva.

Wallerstein I (1990) 'World-systems analysis: the second phase', *Review* **12**.

Walter H (1973) *The Vegetation of the Earth.* English University Press, London.

Ward P M (1976) 'The squatter settlement as slum or housing solution: evidence from Mexico City', *Land Economics* **52**: 330–46.

Ward P M (1990) *Mexico City: the Production and Reproduction of an Urban Environment.* Belhaven Press, London.

Weil C (1983) 'Migration away from landholdings by Bolivian campesinos', *Geographical Review* **73**, 182–97.

West R C and **Augelli J P** (1966) *Middle America: Its Land and Peoples.* Prentice-Hall, Englewood Cliffs, NJ.

Whitmore T C and **Prance G T** (eds) (1987) *Biogeography and Quaternary History of Tropical America.* Clarendon Press, Oxford.

WHO/UNEP (1992) *Urban Air Pollution in Megacities of the World.* Blackwell, Oxford.

Wilkie J, Ochoa E and **Lorrey D E** (1991) *Statistical Abstract of Latin America.* UCLA, Latin American Center Publications, Los Angeles.

World Bank (1984) *World Development Report.* World Bank, New York.

World Bank (1991) *World Development Report 1991.* OUP, Oxford and New York.

World Bank (1993) *World Development Report Investing in Health: World Development Indicators.* World Bank, New York.

WRI (1990) *World Resources Report 1990–91.* WRI, OUP, New York and Oxford.

WRI (1991) *World Resources Report 1991–92.* WRI, OUP, New York and Oxford.

WRI (1992) *World Resources Report 1992–93.* WRI, OUP, New York and Oxford.

WRI (1993) *World Resources Report 1993–94.* WRI, OUP, New York and Oxford.

WRI (1994) *World Resources Report 1994–95.* WRI, OUP, New York and Oxford.

Zisman S (1992) *Mangroves in Belize: Their Characteristics, Use and Conservation.* Natural Resources Institute, Overseas Development Administration, Chatham.

Index

306